T0303885

Systems Engineering Tools and Methods

Engineering and Management Innovations Series

Hamid R. Parsaei and Ali K. Kamrani, Series Advisors
University of Houston, Houston, Texas

Systems Engineering Tools and Methods
Ali K. Kamrani and Maryam Azimi
ISBN: 978-1-4398-0926-6

Optimization in Medicine and Biology
Gino J. Lim and Eva K. Lee
ISBN: 978-0-8493-0563-4

Simulation of Industrial Systems: Discrete Event Simulation Using Excel/VBA
David Elizandro and Hamdy Taha
ISBN: 978-1-4200-6744-6

Facility Logistics: Approaches and Solutions to Next Generation Challenges
Maher Lahmar
ISBN: 978-0-8493-8518-6

Systems Engineering Tools and Methods

Edited by
Ali K. Kamrani • Maryam Azimi

CRC Press
Taylor & Francis Group
Boca Raton   London   New York

CRC Press is an imprint of the
Taylor & Francis Group, an **informa** business

CRC Press
Taylor & Francis Group
6000 Broken Sound Parkway NW, Suite 300
Boca Raton, FL 33487-2742

© 2011 by Taylor and Francis Group, LLC
CRC Press is an imprint of Taylor & Francis Group, an Informa business

No claim to original U.S. Government works

Printed in the United States of America on acid-free paper
10 9 8 7 6 5 4 3 2 1

International Standard Book Number: 978-1-4398-0926-6 (Hardback)

Visit the Taylor & Francis Web site at
http://www.taylorandfrancis.com

and the CRC Press Web site at
http://www.crcpress.com

To my wife and children

Ali K. Kamrani

To my parents

Maryam Azimi

Contents

Preface

Lack of planning and clear identification of requirements has always been the major problem associated with the design and development of any complex system as it results in lack of performance, design failure, and, possibly, expensive design modifications. In this scenario, requirements at the systems level are kept simple in order to integrate new technology, but this often leads to last-minute modifications that impact the schedule and cost. Decisions made at the early stages of the development life cycle will have a significant impact on the overall life cycle, including the cost incurred and the effectiveness of the system. Therefore, a disciplined approach to integrated design and the development of new systems is required. This approach is called systems engineering (SE). In SE, all aspects of development are considered at the initial stages and the inputs derived are used for continuous improvement. SE is "The effective application of scientific and engineering efforts to 1) transform an operational need into a defined system configuration through the top-down iterative process of requirement analysis, functional analysis and allocation, synthesis, design optimization, test, evaluation and validation, 2) integrate related technical parameters and ensure the compatibility of all physical, functional, and program interfaces in a manner that optimizes the total definition and design, and 3) integrate reliability, maintainability, usability, safety, serviceability, disposability to meet cost, schedule, and technical performance objectives."* Figure P.1 shows the scope of the design process for any complex system.

Complex design problems require knowledge from many different disciplines. The team involved in the system development must coordinate during the design process and provide the knowledge required in order to implement the design. During the conceptual phase of the design life cycle, the emphasis is mainly on establishing the requirements and the relationship between them. The need for integration continues when the design moves from the preliminary stages into the detailed designing process. The integrated or concurrent development method requires cross-functional teams, with members from all the functional

---

* *Systems Engineering Fundamentals.* Defense Acquisition University Press, Fort Belvoir, Virginia, 2001.

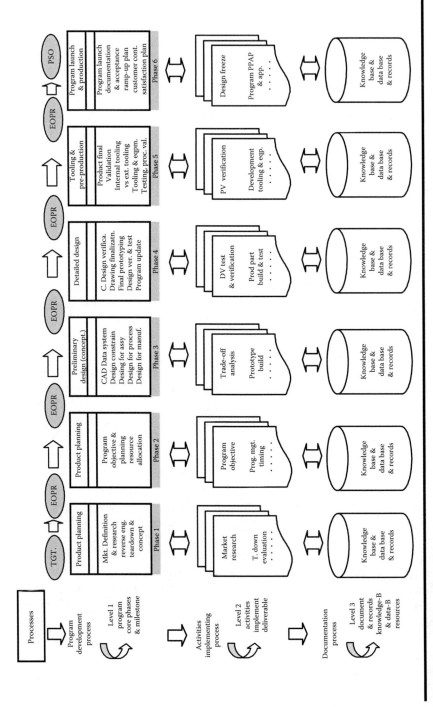

Figure P.1 Scope of SE life cycle.

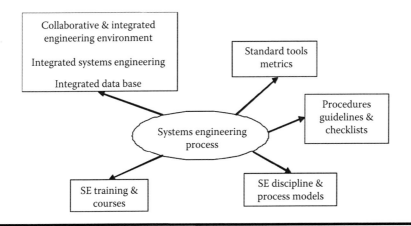

Figure P.2   Integrated SE infrastructure.

areas working closely together, sharing details of their portion of the design as it progresses, and developing all aspects of the system simultaneously. They thus manage the overall life cycle. Concurrent engineering (CE) is defined as the systematic and integrated approach to the systems life cycle design. It is also known as the *design-for-life-cycle* model. CE is the implementation of parallel designs by cross-functional teams, including suppliers. Without empowered team members and the free flow of communication, this method will not function. Figure P.2 provides an overview of the integrated SE infrastructure.

A model that is typically used to define the critical elements of the SE process is the systems engineering capability maturity model (SE-CMM).* These elements are activities that define "what" has to be done. Table P.1 lists these tasks. Tasks for design and development are listed in the engineering group, the project elements provide the management infrastructure, and, finally, the organization elements provide the business infrastructure to support SE efforts.

This book is a collection of methods, frameworks, techniques, and tools for designing, implementing, and managing large-scale systems as presented by experts in this area. It also includes case studies. Sample case studies exhibit the effect of the SE concept and its importance during the design and development of a complex system. This book aims to familiarize the reader with the fundamental issues of SE, laying emphasis on the integrated approach to the design life cycle of complex systems.

Chapter 1 provides a general overview of SE and the engineering design of systems. Several topics like SE, life cycle of SE, and some general approaches to the implementation of SE are discussed and presented. The important relationship

---

* *Systems Engineering Fundamentals.* Defense Acquisition University Press, Fort Belvoir, Virginia, 2001.

Table P.1 SE-CMM

Engineering	Project	Organization
✓ Understand customer needs	✓ Ensure quality	✓ Coordinate with suppliers
✓ Derive and allocate requirements	✓ Manage and control configurations	✓ Define the SE process
✓ Analyze alternative solutions	✓ Manage risks	✓ Manage the system evolution
✓ Evolve the system architecture	✓ Monitor and control technical efforts	✓ Manage the SE support environment
✓ Integrate system	✓ Plan technical efforts	✓ Continuous improvement
✓ Integrate disciplines	✓ Integrate technical efforts	
✓ Testing and acceptance		

between SE and project management is also discussed. Finally, an approach by the U.S. Department of Transportation to address risks and uncertainties is presented at the end of the chapter. The types of risks and uncertainties are summarized and some tools used by systems engineers to reduce and mitigate these risks and uncertainties are presented. Chapter 2 presents evidence of the consensus view that the primary contributors to large and unpredictable increases in cost and schedule of Department of Defense (DoD) acquisition programs due, in part, to shortcomings of both the implementation and execution of SE. Chapter 3 reviews principles of functional analysis, including functional decomposition and functional allocations to physical components, as design activities that establish a system's expected behavior in its particular operating environment. Case studies of the wooden seventeenth-century Swedish warship, the Vasa, and information and communication technologies (ICT) are examined in terms of the services to be provided, the fundamental view of their functional/logical architecture, and, finally, their physical architecture. A classical ICT network is presented and analyzed against a state-of-the art Internet Protocol (IP) multimedia services (IMS) network to illustrate architectural robustness in the fast-changing telecommunication field. The main focus of Chapter 4 is on a series of technical and management activities necessary for the successful VV&T (verification, validation, and testing) of complex systems. This chapter discusses VV&T methodologies and tools for complex systems, which consist of hardware components, software components, and processes. Architectural attributes that affect the VV&T process, verification methods, verification planning and implementation, verification phases throughout the system development cycle, system testing planning and deployment, as well as validation are discussed in detail. A classical virtual network service (VNS) of the late twentieth century is introduced to present an example of VV&T processes. This VNS is then compared to current multimedia networks comprising the next generation networks (NGN). These examples show the evolution of VV&T methodologies and processes in an ever-changing environment. Finally, a case study on the current National Aeronautics and Space Administration (NASA) Constellation Program is presented to illustrate VV&T planning in the aerospace industry. Chapter 5 discusses a framework for assessing technology integration at the systems level. This chapter begins with a discussion of technology forecasting and some common techniques for assessing technology maturity. Then a framework for addressing technology risk and integration issues at the systems level is presented. This discussion is followed by a hypothetical example and concluding remarks. Chapter 6 addresses the development of a business process reengineering (BPR) plan for the restructuring, retraining, and redeployment of elements of the Kennedy Space Center (KSC) workforce during the transition. This case study offers a potential framework of the change process that may be applied to further restructuring at KSC or to the restructuring of other organizations. Case studies provide unique and necessary insights as well as a realistic component of SE knowledge. In Chapter 7, the historical background is presented during the design and development of several large complex systems, spanning three decades from the DoD and NASA. These systems include the F-111 sweep-wing fighter, the large front-loading C-5 mobility aircraft, the serviceable Hubble Space Telescope satellite, the cold war Peacekeeper missile system, the A-10 "tank-buster" aircraft, the stealth B-2 bomber, and the Global Positioning System constellation. Each system provides the backdrop for several SE learning principles, which are depicted in a framework by George Friedman and Andrew Sage. A few principles that are highlighted regularly in the acquisition environment and political context of these systems are technical decision making, requirements definition, logistics support planning, verification and risk mitigation. Chapter 8 proposes an integrated reliability management system for faster time-to-market electronics equipment in the case of distributed manufacturing paradigm. A major advantage of distributed manufacturing is the cost reduction that makes products more competitive. However, the distributed manufacturing process involves many risks in terms of product quality and reliability. Our objective is to address the reliability issue across the product life cycle from equipment design through manufacturing, integration, and field usage. The idea is to bridge the reliability information gap between the manufacturer and the customer through an effective failure-tracking and corrective-action system. Chapter 9 proposes a new integrative approach to the allocation of adjustability and sizing of the mountain bike frame. The case study undertaken deals with a problem of industrial strength with numerous design variables. The underlying platform planning problem is solved utilizing a novel optimization methodology known as Taguchi integrated real-time optimization (TIRO). In order to show the robustness of the model to handle various kinds of market fluctuations, various demand modeling techniques have been detailed and included in the mathematical model. Moreover, the efficacy and supremacy of the proposed TIRO solution procedure has been benchmarked against two sets of test beds that were meticulously generated by using techniques derived from design of experiments (DOE) and various other demand modeling functions. Analysis of variance (ANOVA) is performed to verify the robustness of the proposed solution methodology. The algorithm performance is also compared with three other pure algorithms (GA, AGA, and SGA) where TIRO is seen to outperform significantly in all the problem instances within the test bed. Chapter 10 discusses the importance of manufacturing with regard to designing components of a system. It provides a model for manufacturing under the constraints and conditions of the mass customization environment. The model is based on manufacturing features and entails the concept of the modular design. Chapter 11 discusses various methods and algorithms used in the solution of combinatorial optimization problems. In this chapter, first, a definition of a combinatorial optimization problem is given, followed by a discussion on the depth-first branch-and-bound algorithm, the local search algorithm, and two approaches used for solving these kinds of problems. An introduction to genetic algorithms coupled with an explanation of their role in solving combinatorial optimization problems is then presented. A sample case of the modular design is also presented.

We would like to thank Taylor & Francis Group, CRC Press for publishing this book. Thanks are also due to our friend, the late Ray O'Connell, for giving us the opportunity to fulfill our vision of writing this book. He will be sorely missed.

For MATLAB® and Simulink® product information, please contact

The MathWorks, Inc.
3 Apple Hill Drive
Natick, MA, 01760-2098 USA
Tel: 508-647-7000
Fax: 508-647-7001
E-mail: info@mathworks.com
Web: www.mathworks.com

Editors

Ali K. Kamrani is an associate professor of industrial engineering and director of the industrial engineering graduate program studies. He is also the founding director of the Design and Free Form Fabrication Laboratory at the University of Houston. He has been a visiting professor at the Princess Fatimah Alnijris's Research Chair for Advance Manufacturing Technology (AMT), Industrial Engineering Department at King Saud University, Riyadh. Saudi Arabia. He received his BS in electrical engineering in 1984, his MEng in electrical engineering in 1985, his MEng in computer science and engineering mathematics in 1987, and his PhD in industrial engineering in 1991, all from the University of Louisville, Louisville, Kentucky. His research interests include the fundamental application of systems engineering and its application in the design and development of complex systems. He is the editor in chief for the *International Journal of Collaborative Enterprise* and the *International Journal of Rapid Manufacturing*. He is a professional engineer at the state of Texas.

Maryam Azimi is a PhD candidate in the Department of Industrial Engineering at the University of Houston. Azimi received her BS in textile technology engineering (2001) and her MS in textile management (2004) from Tehran Polytechnic (Amirkabir University), Tehran, Iran. Her current research interests include systems engineering, statistics, and application of data mining in health care and Computer Aided Design/Computer Aided Manufacturing (CAD/CAM).

Contributors

Maryam Azimi
Industrial Engineering Department
University of Houston
Houston, Texas

John Colombi
Department of Systems and
    Engineering Management
Air Force Institute of Technology
Wright-Patterson Air Force Base, Ohio

R. Darin Ellis
Industrial and Manufacturing
    Engineering Department
College of Engineering
Wayne State University
Detroit, Michigan

Gerald W. Evans
Department of Industrial Engineering
University of Louisville
Louisville, Kentucky

Cynthia C. Forgie
Engineering Department
University of Southern Indiana
Evansville, Indiana

Francisca Ramirez Frezzini
Industrial Engineering Department
University of Houston
Houston, Texas

Charles M. Garland
Air Force Center for Systems
    Engineering
Air Force Institute of Technology
Wright-Patterson Air Force Base, Ohio

Tongdan Jin
Ingram School of Engineering
Texas State University
San Marcos, Texas

Ali K. Kamrani
Industrial Engineering Department
University of Houston
Houston, Texas

Madhu Kilari
Department of Business
    Administrations
Texas A&M International University
Laredo, Texas

Nil Kilicay-Ergin
Engineering Division
Pennsylvania State University
School of Graduate Professional Studies
Malvern, Pennsylvania

Gary P. Moynihan
Department of Civil, Construction,
    and Environmental Engineering
The University of Alabama
Tuscaloosa, Alabama

Ricardo L. Pineda
Research Institute for Manufacturing
and Engineering Systems
University of Texas at El Paso
El Paso, Texas

Ravindra Sachan
Industrial Engineering Department
University of Houston
Houston, Texas

Sa'Ed M. Salhieh
The University of Jordan Hospital
Amman, Jordan

Hazem Smadi
Industrial Engineering Department
University of Houston
Houston, Texas

Eric D. Smith
Research Institute for Manufacturing
and Engineering Systems
University of Texas at El Paso
El Paso, Texas

Mukul Tripathi
Department of Mechanical
Engineering
Center for Advanced Manufacturing
and Lean Systems
University of Texas at San Antonio
San Antonio, Texas

Hung-da Wan
Department of Mechanical
Engineering
Center for Advanced Manufacturing
and Lean Systems
University of Texas at San Antonio
San Antonio, Texas

Gary Witus
Turing Associates, Inc.
Ann Arbor, Michigan

and

Industrial and Manufacturing
Engineering Department
College of Engineering
Wayne State University
Detroit, Michigan

Chapter 1

Review of Systems Engineering Scope and Processes

Francisca Ramirez Frezzini,
Ravindra Sachan, and Maryam Azimi

Contents

This chapter provides an introduction in systems engineering process. Several top-ics such as system engineering, life cycle of systems engineering, and some general approaches to the implementation of system engineering are discussed. A discus-sion about the important relationship between system engineering and project management is also presented. Finally, an approach by the U.S. Department of Transportation is presented to address the risks and uncertainties. The types of risks and uncertainties are summarized and some tools used by systems engineers to reduce and mitigate these risks and uncertainties are presented.

1.1 Introduction

A system is a set of objects, with relationships between the objects and their attri-butes (Hall, 1962). The objects here are components of the system. The attributes are the properties of the objects and the causal, logical, and random relationships that connect these components. The environment consists of all the objects that are outside the system. Any change in the attributes of the object affects the system. A system is therefore a collection of parts that interact with each other to function as a whole (Kauffman, 1980). A system is also a set of integrated end products and their enabling products (Martin, 1997).

The word "system" originates from the Greek "systema," which means "an organized whole." The complex networking of resources such as manpower, equip-ment, facility, material, software, hardware, etc. form a system. These resources are components that interact with each other as a system, within or beyond the frame-work of the system. The components are designed to perform many functions, according to the desired objectives and purposes of the system. There are many dif-ferent ways of categorizing systems. These may include natural, man-made, static, dynamic, simple, complex, reactive, passive, precedented, unprecedented, safety-critical, non-safety-critical, highly reliable, non-highly reliable, high-precision, non-high-precision, human-centric, nonhuman-centric, high-durability, and non-high-durability systems. As Magee and de Weck (2002) suggest, "the system can be classified based on the process (Transforming or processing, Logistics and distribution, Trading and storing, Control and regulate) and the outputs (Matter, Energy, Information and Value)" (Magee and de Weck, 2002).

A simple model system has elements such as inputs to the system, processes and mechanisms, constraints, system type depending upon the service/product, and the output. It is very important for the systems engineer to consider the system life cycle.

There is a primary difference between systems engineering and manufacturing. In manufacturing, the product is well defined with a known process, and the main goal of this process is to produce a quality product, while reducing the cost as a measure of performance. In systems engineering this is not true, since the main focus is on defining the problem. The systems engineer adds value only by measuring and reducing risk. A system is a collection of processes performed by the components. As such, a system is a structure with different levels in a complex hierarchical framework. A system can be a new system or an existing system, and is typically composed of many subsystems that are connected by the functional interface.

The stakeholders of the system are its owners, customers, vendors, and every person directly or indirectly related to the system. The objectives are the requirements and the combination of the goal, based on the defined needs. The main objective of the system should be to achieve its stakeholders' concern. The design of methods and models are the two factors in systems engineering. The integration of these methods and models within this system and with other systems decides the complexity of the system, which in turn describes the complexity of the systems engineer's task. The integration is basically the networking of the detailed breakdown of the different subsystems, components, or processes. The two main functions of systems engineering is the verification and validation of the elements of the system in order to achieve the specified quality standards or based on a certain level of acceptance.

The design involves answering the questions of type: what, how, and when to define the stakeholders' requirements. As is well known in the systems and software engineering literature (Boehm, 1981; Haskins et al., 2004), the earlier one finds a flaw in the design, the fewer resources are required to fix the error. The cost associated with the system life cycle is research and development (R&D), operation, maintenance, and support system. System performance is measured by aspects such as availability, reliability, maintainability, quality, disposability, and supportability.

1.2 Need for Systems Engineering

The current design environment is dynamic in nature, with changing requirements due to changes and advances in technology. This makes the design of any system complex. Globalization has also increased competition, forcing companies to decrease the cost and development time.

The ambiguity in defining the requirements and the lack of proper planning are the major factors that drive the need for a systems engineering interference. "A system can be generally defined into four phase i.e. conceptual design, preliminary design, detailed design and development and operation and management. Current methods focus mainly on the last two phases. The cost allotted to these two phases is comparatively more; approximately 75% of the total project cost" (Smith, 2006). But systems engineering focuses on spending more percentage of funds and resources on the initial concept design and preliminary architect, which can reduce the cost in the next two phases—not in a direct manner, but by improving the quality of the design, improving the relation between components and processes, reducing errors and defects, reducing the development time, and finally improving the overall performance of the system. An important function of systems engineering is to balance the cost and performance of the system.

"The problems and challenge with many project is either many project are lagging behind the schedule, running over the estimated cost or are degraded into the functionality as per the original plan" (Gonzalez, 2002). Systems engineering adds value to the development, fielding, and upgrade of a system. The engineering of the system is the design and integration with broad perspective and clear system requirements. "The things which a system engineer focus on is identify and evaluate the requirements, alternatives, risk management and uncertainties, design house of quality for the system under consideration, and to manage the activities of the project" (Gonzalez, 2002). In the identification phase, the main aim and goal of the systems engineer is to determine the different trade-offs and then proceed with the design. The second phase is to analyze the trade-offs, and then analyze and formulate a methodology technique to deal with these trade-offs. The final phase is to decide on the best choice, based on three aspects that decide the feasibility of the project.

The first aspect is to decide whether or not the project or system is technically feasible (Gonzalez, 2002). These decisions are based on the technical background and domain of a specific area. It is necessary to decide on two or more alternatives by analyzing the technical trade-off, based on the requirements, availability, and capability of the option. The decision should optimally be based on utilization and capacity effectiveness; only then is it labeled as feasible. The best technical solution may not always be the best solution, as this solution will be based on the technical aspects only, but will not consider the cost involved and the time constraint (Gonzalez, 2002).

The second aspect of a system is the cost. Before starting any project, the cost associated with it is estimated. The most important goal of the systems engineer will be to stay within the estimated limit of the total cost involved. The different alternatives should be evaluated on the basis of the life cycle cost (Gonzalez, 2002), such as the cost involved in the initial development of the system, the cost associated with the operation and maintenance of the system, and other supply-related costs. The project which is technically the best may not be the best in cost-effectiveness; in such cases, the choice of the better option is the one which meets both technical feasibility and cost feasibility.

The third most important (and often neglected) aspect is the time constraint that is associated with the schedule-related issues of the project. Most projects that technically sound good lag behind schedule due to budgetary constraints that make the project unsuccessful. There are many reasons for possible delays, such as employee incompetency in a particular skill, or the absenteeism or shortage of resources; often, there are delays due to internal or external activities or process which are directly or indirectly related to the system's need and requirements. For example, a part is waiting for assembly with another part that is being outsourced, as it was difficult or costly to manufacture that part in-house.

There are uncertainties in evaluating and selecting the best alternative, due to the unpredictability of increasingly complex systems and their requirements. It is very difficult to predict accuracy in a precise manner, which generates a risk factor in the design during the systems engineering process. The risk that is associated with the life cycle design is managed through identification, analysis, and the mitigation of uncertainties. The first step is to identify the different potential issues that affect the performance and progress of the system. After identifying the problems, the next step is to distinguish the issues on the basis of uncertainty and importance. The problems are then prioritized on the basis of uncertainty and importance, and are ranked from high to low. After categorizing, the issues are mitigated depending on their concern, importance, and uncertainties. "Identifying the potential issues is based on questions which will focus on the areas like requirement, concept of operation, need analysis, system design, the complexity of the design, technicalities involved, the method of integration, cost, capacity, resources, design and concept interface, integration of different process, quality, flow, operation and maintenance, test and validation" (Gonzalez, 2002). Mitigation involves the application of different theories and use of various tools, such as simulation, optimization, queuing theory, market research, forecasting techniques, trade-off studies, all of which are supported by surveys and questionnaires.

1.3 Project Management and Systems Engineering

Project management deals with management issues; in contrast, systems engineering deals with technical issues (Gonzalez, 2002). The project management team is headed by the project manager, but the systems engineering team is led by a skilled and experienced systems engineer who has knowledge in the required technical field. The main aim of the project manager is the success of the project, while the main aim of the systems engineer is the success of the system. The main goal of the systems engineer is to transform the functional requirements into operational needs, and to set the performance parameters to design the system configuration. Research has proved that many of the unsuccessful projects, when tracked back, mainly failed due to the lack of effective systems engineering. The complexity of the design forces the systems engineer to divide the system in subsystems for development, which in turn leads to the formation of a complex subsystem structure. This complex framework builds the different sub-products. At the assembly stage, these levels of decomposition should match, which is only possible if the subsystems are as per the required specifications. However, because of such a complex structure, it is difficult to provide the necessary accuracy. Systems engineering defines the requirements and specifications for the subsystems, their design, and the process for their development, and then integrates the subsystems in order to build the whole architecture. Finally, testing validates and verifies the process.

Systems engineering involves the following functions: locate the requirements, and identify the problem from previous solutions; study and analyze the problem and possible solutions; develop a conceptual model; design and formulate a physical or virtual model based on the requirements and available resources; create a testing procedure with knowledge interference; analyze the various available options to resolve the current issue; and pay attention to feasibility in relation to technicality, cost, and time constraints. The overall discussion is about how to see the big picture of the system and to apply the appropriate technical knowledge along with proper coordination, which can be achieved by the application of project management strategies.

Successful project management is defined based on how complete and comprehensive the project is, the availability of required resources, and the interconnection and precedence of tasks that define the interdependence. This interdependence decides the complexity of the system. The detailed process plan for completion of each single process should be defined, along with the method. The coordination of the instruction and commands should be circulated properly without any filtration. As the information percolates down through the vertical hierarchy of the system architecture, there might be the loss of information, which could be harmful for system performance and should be avoided by identifying and closing the loopholes in the system. This is all about planning and structuring the task.

The next important function of project management is to keep track of intricate details of the project, and to develop a procedure to keep updating the whole information database, so that the current status of the system is available whenever need. This is achieved by the information technology advancement. The integration of software and hardware is the core function of the systems engineer. Everything revolves around the interfacing of the different component systems, so that they can communicate effectively within the given frame of reference.

The most important task of project management in systems engineering is the feedback that helps to generate and predict any uncertainties and flaws. It alerts the systems engineer to be proactive and to be prepared for any issues. The reactive process may be useful, but it needs an experienced and knowledgeable person who has the capability to make decisions. This is not always possible, so the feedback tool will always help to track the performance of a process and help to gather data for improving the performance, whenever required. The discussion above can be summarized into four stages: track, measure, feedback, and plan again after removing errors.

The system engineer definition falls into four categories:

1. System engineering is a function involving technical activities.
2. System engineering is a function involving technical and management.
3. System engineering is a function involving technical activities which can be performed by any one from technical or scientific discipline.
4. System engineering is a function involving technical and management activities which can be performed by any one from technical or scientific discipline (Gonzalez, 2002).

Four main systems engineering activities are considered for the success of any large-scale projects. "These are: identify and evaluate alternatives, manage uncertainty and risk in our systems, design quality into system and handle program management issues that arise" (Gonzalez, 2002). The first activity (feasibility analysis) is the most important, as this decides the probability of success. The feasibility of all the alternatives should be measured. There are three feasibility measures: technical, cost, and schedule. Technical feasibility checks the construction, operation, and maintenance of the system, as well as the available resources. Cost feasibility checks whether the system can be built, operated, and maintained within estimated or available cost. Schedule feasibility checks the time constraints.

The second activity is managing risk and uncertainty. This is also a very important aspect because it helps systems engineers to avoid any errors. The three important aspects of risk management include identification, analysis, and mitigation.

The third activity is designing quality into systems. The main focus in this activity is to deal with the factors that negatively impact quality. The factors affecting quality are complexity, inflexibility, and lack of standardized components, reliability, and availability. Systems that are complex and intricate are hard to maintain. Inflexible systems are not easy to adapt, and non-standardized components are difficult to maintain. An unreliable system is non-functional and prone to defects and faults.

The fourth activity is dealing with project management issues. The main concern of this activity is to deal with the three important aspects of a good project: complete, comprehensive, and communicated. The impact of systems engineering on project implementation is due to conception, requirement analysis, design, implementation, testing, system acceptance and operation and maintenance, project planning and control, technical planning, technical audits and reviews, and cost estimates.

Systems engineering is the process of transforming ideas into a function to add value to a product even with identified cost and resource constraints. Systems engineering can be defined as the application of technological and management tools to attain the defined objective by meeting the specification of the desired performance. This also involves requirement analysis, function analysis and allocation, design, simulation of design, and then optimizing the process in order to achieve the required objective within the given specifications. According to the Department of Defense (DoD), "System engineering is Management of technology where Management involves the interaction of an organization with the environment and the technology is the result of the organization, application and the delivery of scientific knowledge" (DoD Systems Engineering, 2010).

As stated previously, systems engineering is the concept that deals with the design of complex systems, and views the big picture of the general concept, which is defined based on the requirements to create a vision for the system and its implementation. It applies management and technical tools to analyze and resolve issues, and helps to create a formal structure for the system. Systems engineering is not just a set of tools, it is a complete process. Systems engineers must make a better choice to do things by building a unique framework of methods to handle a particular task, considering all other factors related to it, and foreseeing other opportunities with respect to other methodologies for performing the same task. It automatically incorporates the phase that analyzes and eliminates inferior choices. Basically, systems engineering provides the platform to make comparison between the few best opportunities by using methodologies and available tools such as simulation, optimization theory, and queuing theory.

For example, the simulation method is classified as soft, hard, and rapid (Buede, 2009). Soft simulation can help to simulate the entire process by using current or past data and on the basis of descriptive statistics. This simulation model is formulated by using and fitting the data available within some statistical distribution that can be used and analyzed to study the pattern. The pattern can be used to locate current flaws in the process (i.e., the bottleneck process of redundant operations), or it can be used to predict the future process and pattern, based on the particular distribution process. The second approach to simulation is to simulate a particular system by making a replica, and then studying the performance and features by using computer-aided engineering (CAE). Virtual reality is the most modern method of simulating a process and a product. These simulation techniques are an integral part of systems engineering, and have been tested in many cases, along with optimization and other analytical methodologies.

> Systems engineering definition is based on structure, function, purpose, knowledge and working. The structural model deals with problem formulation, analysis of the formulated problem and the interpretation of the results. The functional models require the system engineer to have the necessary knowledge of tool and techniques available for designing the system framework along with the function module as per the requirements and objective. The knowledge model relates to the information and knowledge integration in the system for the purpose of continuous

improvement. A working model of system engineering is the application of proven standards, procedures, tools for planning organization, control, execution and establishment of system requirements, design, management, fabrication, integration, testing and integrated logistic support, all within the required optimum cost and schedule (Buede, 2009).

How does systems engineering deal with a complex system? Various engineering branches keep searching for the answer. The main reason why most large complex software projects fail is because they do not address the defined requirements in a systematic way. The complexity of systems is also impacted if the user requirements are not clearly defined. To overcome this problem, the complexity for systems must be adequately addressed in the requirements definition for the system so that a feasible solution approach can be formulated. The requirements definition falls within the systems engineering domain, and is considered to be the major part of the system life cycle. The approach should be efficient and flexible in order to provide the level of robustness that allows each of the domain experts to define the requirement and achieve the desired end objectives. This could reduce the level of complexity in the system and its design.

The life cycle of any system is defined between the problem/concept definitions phase (conception phase) until system disposal. "The different stages within system life cycle consist of the conceptualization of idea, define the requirements, analyzing the needs using system engineering tools, construct a system framework, theoretical and practical validation of the system, site implementation, integration and testing followed by the validation and acceptance of the system" (Gonzalez, 2002).

The conception phase is followed by the identification of the requirements and how these can be achieved within the scope of the different available alternatives. Compared to the conception phase, a more detailed analysis is done in the identification phase. This analysis involves breaking down and differentiating the needs into two categories: high-level and detailed.

The next stage is the design of a system framework, the implementation phase. The architecture is defined using different systems engineering tools, which also involves decision making regarding how the overall system will be integrated. The implementation phase is a transition from the design phase to the product phase. The implementation phase must be integrated with testing for the required validation and verification. Acceptance testing of the system's performance for its operational capabilities is required by the customer. This phase also includes training the user to operate the system.

The systems engineer must be involved in all phases of the system life cycle. Systems engineering is not just application of tools, but is a quality improvement process composed of technical and management theories to develop a process for analyzing and resolving system-related problems.

"The life cycle for any system can be generalized into

1. Identification and conceptualization of the requirements based on stake holder's needs.
2. Analyze the requirements and categorize the needs into the components and process and decide the specifications. The functions of the component should be complete, correct and detailed.
3. Design a general system for obtaining the primary needs using different system engineering tools.
4. Design the details by dividing the system into subsystem with the required specification. The framework should be build with the clearly defined interface.
5. Implementation." (Gonzalez, 2002).

1.3.1 Systems Engineering Functions and Process

The two important functions of systems engineering are design and decision making. The distinguishing between a good decision and a good outcome is important. Decisions have to be made with the best information available at the time, realizing that the outcomes associated with the decision remain uncertain when the decision is made. The type of questions asked during different decision-making steps are as follows (Farrell, 2005):

1. *Conceptual design:* Is the conceptual design idea feasible? Which technologies or what combination of technology should be used? Which technologies should be used for subsystems? What are the different possibilities for extended research?
2. *Preliminary design:* Is the preliminary design feasible and effective? Which alternatives for the primary system architecture are optimum and appropriate? What are the resource allocations for different components, depending upon different function, prototype, and requirement? Has the preliminary design been validated?
3. *Full-scale design:* Is the full-scale design feasible? What are the costs and benefits of manufacturing a product or product component against purchasing it ("make or buy" decisions)? What is the desired functionality from the different available alternatives?
4. *Integration and qualification:* Have the different components been satisfactorily integrated? Has the integration process been successful in terms of schedule, time and cost constraints, equipment, people, facility, and resources? Which models can optimize the overall system design? What are the current and prospective issues? Which criteria have been used to measure performance?
5. *Product refinement:* What is the time frame for product improvement? Which technologies are required, or which technologies need advancement? What are the cost issues?

The systems engineering process is about trade-off and integration. The systems engineer need not be an expert in all technical domains, but should have sufficient knowledge about potential technology, available methodologies, and should have management skills. The systems engineering team consists of people who are involved in the design and development process, and who have many different specialties. Engineers with knowledge of the technologies associated with the system's concept are needed to provide the expertise needed for design and integration decisions throughout the development phase. The team not only consists of members from traditional engineering fields such as electrical, mechanical, and civil engineering, but also from other disciplines such as social sciences, so as to collectively address psychological, informational, physical, and cultural issues. In addition, engineers who model and estimate system-level parameters such as cost and reliability are also needed. Analysts skilled in modeling and simulation—more and more of which is done on the computer, rather than with scaled-down mock-ups of the system—are also important members of this team. Managers that are in charge of meeting cost and schedule milestones need to be part of this team.

The systems engineering process is divided into the systems engineering technical process and the systems engineering management process. The systems engineering management process is composed of activities, development phasing, life cycle integration, and the systems engineering process. The systems engineering technical process defines the major activities within the system. The different elements of the systems engineering technical process are process analysis, requirement analysis, functional analysis and allocation, design synthesis and process output, system analysis and control, and finally verification.

The systems engineering checklist is as follows:

Do you understand the systems engineering process?

Are you implementing an optimal systems engineering process?

Have you implemented proper and sufficient systems engineering controls and techniques?

Are you implementing systems engineering across the whole development life cycle?

Is there an experienced and skilled systems engineer directing the systems engineering effort?

Is a system engineering representative providing input to or comments on all product change proposals?

Is the systems engineer seeing that all the various development efforts are coordinated and integrated?

Do you know what software development life cycle your project will be employing and how it coordinates with the software and project life cycles?

Are you considering all phases of the entire life cycle in your requirements, architectures, and designs?

Are you implementing an integrated product environment?

Have you established integrated (interdisciplinary) product teams?

Have you included all the necessary disciplines on the integrated product teams?

Are you documenting all studies, decisions, and configurations?

Have all internal and external interfaces been defined?

Are all your requirements verifiable?

Do all your requirements trace to products and vice versa? (International Council on Systems Engineering).

1.3.2 Systems Engineering Management Plan

One fundamental part of systems engineering is planning systems engineering activities, what is done, who does it, how it is to be done, and when the activities are expected to be completed. The purpose of the Systems Engineering Management Plan (SEMP) is to document the results of the planning process.

An SEMP is basically a planning document that should be addressed by the systems engineer at the end of the conception phase, and should be updated as needed thereafter. It primarily schedules activities and reviews, and assigns SE functions. The level of detail necessary depends on the size of the team, scope of the project, and some other aspects.

The systems engineer will assign the person who is responsible for and who will document the mission objectives(s), architecture/design, verification and validation, requirements and hierarchy, configuration control (when documents are placed under formal control, archived, and assigned a method of distributed) and management, verification activities and tracking, interfaces and interface control documents (ICDs), mission-specific environment levels and limits, resource budgets, risk management, and acceptable risk (Beale and Bonometti).

1.3.3 Work Breakdown Structure

Work Breakdown Structure (WBS) is a tool for scheduling and tracking tasks. A WBS is a hierarchical breakdown of the project work, which can also include the deliverable items and services. It divides the project into manageable tasks; next, responsibility for accomplishing each of the tasks can be assigned. A WBS can be represented in a hierarchical tree form; on the bottom level of each branch of the WBS is the task, and (optionally) the person assigned to that task. WBSs are very useful during the implementation phase of a project.

A WBS can also be represented by a Gantt chart, which represents a list of the tasks in a numbered outline hierarchy in labeled columns. The responsible person for each task is listed in the columns. The tasks are on the bottom level of the entries in the "WBS" column, with names of the individuals who are responsible for that task (Beale and Bonometti).

1.4 Approaches for Implementing Systems Engineering

1.4.1 Traditional Top-Down Systems Engineering

Traditional top-down systems engineering (TTDSE) is a systems engineering process that begins with an analysis of what problem needs to be resolved. This is usually done with an analysis of the current meta-system (the system of interest) performing one or more missions for the primary stakeholder. This layered process can have as many layers as needed; the bottom layer addresses the configuration items (CIs) that the discipline engineers will design.

Once the CIs have been designed and delivered for integration, the verification, validation, and acceptance testing process begins. Each layer at the decomposition process are verify against the associated derived requirements. At the system level, validation against the concept of operation and acceptance testing is conducted (Buede, 2009).

1.4.2 Waterfall Model of Software Engineering

This is a model that is applied in software engineering, and is characterized by the sequential evolution of typical life-cycle phases described previously that allow iteration between adjacent stages. The major problem with the waterfall model is that it could frequently occur in iteration between phases that are not adjacent (Buede, 2009).

1.4.3 Spiral Model

The spiral model addresses the need to shorten the time period between the users' statement of requirements and the production of a useful product with which the users could interact.

This model has four major processes, starting in the top left and moving clockwise: design, evaluation and risk analysis, development and testing, and planning with stakeholder interaction and approval. The radial distance to any point on the spiral is directly proportional to the development cost at that point. The spiral model views requirements as objects that need to be discovered, thus putting requirements development in the last of the four phases as part of planning. The early emphasis is on the identification of objectives, constrains, and alternate design. These objectives and constrains become the basis for the requirements in the fourth step. The number of iterations around the spiral is variable, and is defined by the software or systems engineers. The final cycle integrates the stakeholders' needs into a tested and operational product (Buede, 2009).

1.4.4 Object-Oriented Design

Object-oriented design, also known as OO design, is a bottom-up process that begins by defining a set of objects that need to be part of the system in order to achieve the desired system-level functionality. Inheritance and information hiding are two key aspects of OO design. Inheritance occurs after the addition of specialization to the general object; this now specialized object "inherits" all the properties (methods and data) not overridden by the specialization. This process is also known as modularity (Buede, 2009).

1.5 Adapting the Systems Engineering Paradigm to System-of-System Programs

System of Systems (SoS) is basically defined as a large and complex system made up of large-scale systems designed to work together. In order to reduce the complexity involved with these SoS programs, a systematic approach is required. Systems engineering will help in architecture, design development, and in sustaining these SoSs. SoS encompasses integration of the legacy system (existing system that is upgraded through integration of evolutionary technologies) and an entirely new system. The main objective of SoS development is to optimize the integration of independent systems that have the potential capability to execute in collaboration, so as to satisfy a set of mission needs. Various SoSs such as the future combat system (FCS) and the integrated deepwater system (IDS) are examples of systems integrated with control command, communication, and computers in order to improve and optimize the performance of the overall system (DoD Systems Engineering, 2010).

1.6 Addressing Uncertainty and Risks in Systems Engineering

1.6.1 Types of Uncertainty and Risks

One of the key factors being analyzed to measure the degree of success in a systems engineering project is the ability of the system to manage and overcome uncertainty and risks. According to the U.S. Department of Transportation, some of the factors that cause uncertainty and risk in projects are (Mitretek Systems Inc., 2002)

Complexity
Processing capacity/communication bandwidth
Rate of technological change
Information imperfections

Complexity: The more parts a system has, the more complex it becomes. Complexity in a system brings uncertainty in two ways. First, complex systems are really difficult to visualize and understand, which translates to the fact that the chances of effectively satisfying the requirements could decrease with the complexity of the system Second, it could be difficult to define and maintain interactions and interfaces within the system when we are changing any of its parts while building on it in the field. Changes in complex systems could lead to undesirable results.

Processing capacity/communication bandwidth: When we build a system, we may be uncertain about the capacity required to process the data we want to process. This uncertainty can generate from one of three possible scenarios. In the first scenario, we calculate and estimate the capacity correctly, in which case the system will handle the data satisfactorily. In the second scenario, there is an overestimation of the capacity required, and the system will not handle the data satisfactorily. Third, there is an underestimation of the processing capacity, and it is impossible for the system to handle the actual volume. The second scenario is a problem when the cost of the system is the main priority, and the third outcome is clearly a problem. This concept can be translated to computer hardware as well as to humans.

Rate of technological change: Nowadays, changes in software technologies occur at a really fast pace. Given the rapid change in technical capability, projects can face uncertainty in choosing the right technology. Also, managers have to make choices about which technology to deploy if they are going to implement systems or make schedules.

Information imperfection: These are deficiencies in what we know that make it difficult to make a choice with which we are comfortable. These imperfections can result from either the quantity of information or from its quality. The goal in systems engineering is to determine whether there are information imperfections, and to improve on the existing information.

1.6.2 Systems Engineering Tools for Addressing Uncertainty and Risks

The U.S. Department of Transportation considers seven tools to address uncertainty and risks (Mitretek Systems Inc., 2002):

1. Project scheduling and tracking tools
2. Trade-off studies
3. Reviews and audits
4. Modeling and simulation
5. Prototyping
6. Benchmarking
7. Technical performance measures

Project scheduling and tracking tools: Three basic scheduling and tracking tools are considered: Gantt charts, activity networks, and WBSs (described in Section 1.3.2). An activity network shows tasks and the order in which they should be executed.

This type of activity network is usually used to show the critical path, which is the sequence of activities that tends to delay the culmination of a project. Project managers have to monitor this critical path more closely than they do other activities. However, this critical path can unexpectedly change, which is why the project manager should verify and continuously monitor network activities.

Trade-off studies: This is one of the ways to deal with information imperfections and technological change. Trade-off studies are often conducted as paper studies, where the analyst assesses product solutions using the technical descriptions from the vendors. The analyst constructs a map that assesses the technical descriptions of the products against the requirements. Another way to perform trade-off studies is by testing product solutions in a controlled environment that simulates the actual conditions of the system as closely as possible. The implementation of this type of trade-off studies is not always possible, but it usually gives the most reliable results.

Trade-off studies can also be used to address technological change. The studies are capable of forecasting when technical advances are going to migrate from the laboratory to the marketplace. This is used when a decision about acquiring new technology can wait, or when system implementation is in its final stages.

Reviews and audits: Design reviews are useful to ensure that the system design meets the requirements and needs of the end user. Another goal of these reviews is to ensure that the system design is understood. It is also used to find flaws in the design, which can be corrected before the implementation. It is also really important to communicate the evolution of these reviews and audits to all interested parties.

Modeling and simulation: This approach makes more sense for systems that have not been built. Modeling and simulation helps to examine potential bottlenecks or impediments to the performance of the system. Once the potential problem areas are identified, the design can be modified in order to reduce errors and negative impacts in the system. These tools are really useful in reducing system performance uncertainty.

Prototyping: Prototypes are operational models of the system, but usually in a scaled-down format. They can be used to create a version of the system being used for benchmarking system performance. Prototypes can also be used to evaluate whether it is possible to interface elements of a system that have never been linked together before, helping to reduce uncertainty related to interfaces.

Benchmarking: Benchmarking is useful in measuring the actual performance of a system, and ensures that it can meet the requirements. Usually, benchmarking in used on an implemented version of a system, as part of acceptance testing. It can also be used against scaled-down versions of the system to measure performance of the scaled-down processes, usually to study if the increase of a system will not significantly change its performance.

Technical performance measures: These are characteristics of a system that impact its performance. These measures should be established early on in the project. Subsequently, as the project evolves, the system performance measures achieved at any stage are compared to the design goals. Modifications and improvements should be implemented at this stage to achieve the defined goal.

Authors

Francisca Ramirez Frezzini received her bachelor's degree in materials engineering specializing in polymers from Universidad Simon Bolivar, Caracas, Venezuela. Her thesis research was on the topic of "statistic approximation in the formulation of polypropylene, with several additives." She has over three years of experience in research and development, technical management, and quality. She received her master's degree in industrial engineering from University of Houston, Houston, Texas.

Ravindra Sachan is pursuing his master's degree in industrial engineering at the University of Houston, Houston, Texas. He is currently conducting research on the implementation of "systems engineering in health care." As part of his MS curriculum, he has conducted research in various fields, including systems engineering, lean manufacturing, simulation, optimization, and CAD/CAM. While pursuing his undergraduate studies in production engineering in India, he underwent training in Siemens Ltd. and Mukand Ltd. He has 35 months of work experience in such diverse fields as design, product development, and process planning and manufacturing, and has successfully accomplished and implemented several projects in companies such as Godrej and Boyce Mfg. Ltd., ThyssenKrupp Elevators India Pvt. Ltd., and Bharat Gears Ltd. He served as an academic secretary in his junior days in college. As a graduate student, he was a member of the Mechanical and Production Engineering Student Association.

Maryam Azimi is a PhD candidate in the Department of Industrial Engineering at the University of Houston. Azimi received her BS in textile technology engineering (2001) and her MS in textile management (2004) from Tehran Polytechnic (Amirkabir University), Tehran, Iran. Her current research interests include systems engineering, statistics, and application of data mining in health care and CAD/CAM.

References

Beale, D. and Bonometti, J. Fundamentals of lunar and systems engineering for senior project teams, with application to a lunar excavator. ESMD Course Material. <http://education. ksc.nasa.gov/esmdspacegrant/LunarRegolithExcavatorCourse/Chapter2.htm>

Boehm, B.W. 1981. *Software Engineering Economics*. Englewood Cliffs, NJ: Prentice-Hall.

Buede, D.M. 2009. *Engineering Design of Systems Models and Methods*. Hoboken, NJ: John Wiley & Sons.

DoD Systems Engineering. 2010. ACQWeb—Office of the Under Secretary of Defense for Acquisition, Technology and Logistics. <http://www.acq.osd.mil/sse/> (February 25, 2010).

Farrell, C.E. 2005. Adapting the systems engineering paradigm to system-of-systems programs. In *Proceeding of 43rd AIAA Aerospace Sciences Meeting and Exhibit*, Reno, NV.

Gonzalez, P.J. 2002. *Building Quality Intelligent Transportation Systems through Systems Engineering*. Rep. No. FHWA-OP-02-046. Washington, DC: Intelligent Transportation Systems Joint Program Office US Department of Transportation.

Hall, A.D. 1962. *A Methodology for Systems Engineering*. New York: Van Nostrand.

Haskins, B., Stecklein, J., Brandon, D., Moroney, G., Lovell, R., and Dabney, J. 2004. Error cost escalation through the project life cycle. In *Proceedings of the INCOSE Symposium*, Toulouse, France.

Kauffman, D.L. Jr. 1980. *Systems One: An Introduction to Systems Thinking*. Minneapolis, MN: Future Systems Inc.

Magee, C.L. and de Weck, O.L. 2002. An attempt at complex system classification. In *2002 MIT ESD Internal Symposium*, University Park Hotel, Cambridge, MA. <http://esd.mit.edu/WPS/ESD Internal Symposium Docs/ESD-WP-2003-01.02-ESD Internal Symposium.pdf>

Martin, J.N. 1997. *Systems Engineering Guidebook: A Process for Developing Systems and Products*. New York: CRC Press.

Mitretek Systems, Inc. 2002. *Building Quality: Intelligent Transportation Systems through Systems Engineering*. Washington, DC: Intelligent Transportation Systems Joint Program Office US Department of Transportation.

Smith, L.D. 2006. KBArchitecting*—Knowledge based architecting for MORE EFFECTIVE SYSTEMS. Forell Enterprises, Inc., Buena Park, CA. <http://www.eforell.com/papers/KBArchitecting%20White%20Paper%2011Aug2006.pdf>

Chapter 2

Current Challenges in DoD Systems Engineering

Gary Witus and R. Darin Ellis

Contents

Department of Defense (DoD) acquisition programs have a history of problems with cost growth and schedule slip. Recent emphasis on revitalization and application of systems engineering (SE), with a special focus on early phases of the system's development life cycle, is aimed at reducing these risks. Revitalization of SE alone, though, will not result in the full potential of desired improvements. While SE methods, tools, and practices have advanced, there are still problems with the implementation, integration, and execution of the SE functions within the acquisition community. This chapter describes some of these issues with reference to recent case examples including the future combat system (FCS).

Despite notable successes, Department of Defense (DoD) acquisition programs continue to be plagued by large and unpredictable growth in cost and schedule, with the accompanying risk of program cancellation. In an article titled *Revitalizing Systems Engineering in DoD* (Wynne and Schaeffer, 2005), the Under Secretary of Defense for Acquisition, Technology, and Logistics wrote

> Our studies show that in cases where programs were started with requirements that exceeded resources, costs increased from 55 percent to nearly 200 percent, and schedule delays jumped an estimated 25 percent. *Early application of systems engineering* will give DoD's top decision makers the necessary confidence in a program's ability to define and match technical requirements with resources—in other words, to stay on budget and on schedule—and to define, understand, and manage program risk (Emphasis added).

While deficient systems engineering (SE) contributes to cost and schedule growth and unpredictability, there are other significant root causes that are external to SE. These external problems both limit the benefit that can be achieved by revitalized SE and undermine effective SE. SE deficiencies in DoD acquisition programs are, to a large extent, issues of organization, management, incentives, subject matter expertise, and SE experience. These are more problems of *implementation, integration, and execution* of SE in the acquisition programs than of SE methods, tools, and procedures. These problems have led to broad recognition that there is significant dissatisfaction with DoD acquisition performance (see Table 2.1).

There are several primary contributors to large and unpredictable growth in cost and schedule of DoD acquisition programs. The most fundamental issue with the implementation of SE is due to overoptimistic estimates and a lack of clear lines of ownership and accountability. Unrealistic, incomplete, pretend, and conflicting requirements are external inputs to the program that cannot be addressed by SE but undermine SE execution and confidence in SE. Convergent programming, in which one program is executed under the assumption that another program will provide a critical element, is a major source of risk. Convergent programs are structured to make use of a common infrastructure to achieve interoperability among users and reduced total logistics burden (e.g., the joint tactical radio system). Even with realistic estimates, there are often problems with execution, including requirements management. Finally, SE is facing a whole new series of challenges with complex systems-of-systems (SoS) architectures and spiral development and

Table 2.1 Dissatisfaction with DoD Acquisition Performance

"… The committee is concerned that the current Defense Acquisition Framework is not appropriately developing realistic and achievable requirements within integrated architectures for major weapons systems based on current technology, forecasted schedules and available funding …" (House Armed Services Committee, May 2005)
"There is growing and deep concern within the Congress and within the Department of Defense (DoD) Leadership Team about the DoD acquisition process. Many programs continue to increase in cost and schedule even after multiple studies and recommendations that span the past 15 years. In addition, the DoD Inspector General has recently raised various acquisition management shortcomings. By this memo, I am authorizing an integrated acquisition assessment to consider every aspect of acquisition, including requirements, organization, legal foundations (like Goldwater-Nichols), decision methodology, oversight, checks and balances—every aspect." (Secretary of Defense, Gordon England, 2005)
"Both Congress and the Department of Defense senior leadership have lost confidence in the capability of the Acquisition System to determine what needs to be procured or to predict with any degree of accuracy what things will cost, when they will be delivered, or how they will perform." (Kadish et al., 2006)
"It has almost become a cliché to state that the numerous reforms initiated over the years have not had the desired effect and that today DoD faces the same acquisition problems. The proposals to correct problems have run the gamut of adding controls, increasing management layers, streamlining, and decentralizing. They have often sought coercive, procedural, and organizational solutions to make things happen without necessarily addressing why they were not happening previously. For example, there have been recommendations aimed at improving the realism of cost estimates, but these are difficult to implement since the acquisition process itself does not reward realism." (Fox et al., 1994)

evolutionary acquisition (EA) processes. In spiral development, the requirements and design coevolve. While this may be a practical approach to keep expectations, goals, and technology maturity/risk in harmony, the cost, schedule, and ultimate system performance are difficult to predict and control. A further technical challenge of developing complex, adaptive systems with the potential for unforeseen positive and negative indirect feedback is the emergence of unforeseen properties. Detecting, disabling, exploiting, and designing emergent properties are major technical challenges for complex adaptive SoS.

The objective of this chapter is to explore the root cause and context of factors that play a large role in limiting the effectiveness of SE in DoD acquisition: First, we address the *implementation and integration* of SE in large-scale acquisition programs. Next, we address challenges in the *execution* of SE in the acquisition program. Finally, we consider SE challenges in the development of complex, adaptive systems, or "SoS."

2.1 Implementation and Integration Issues

Programs are often initially oversold with optimistic cost, schedule, and technical maturity estimates. This is the consequence of the "moral hazard" in the procurement process in which a program with realistic cost, schedule, and technical maturity estimates has to compete for funds against programs with optimistic or "success-oriented" assessments. This creates pressure to forecast through rose-tinted glasses. There are inherent conflicts and ambiguities in the position of the acquisition manager with respect to the senior officers of their service branch that encourage overoptimistic assessments. The promotion and future assignments of an acquisition manager are subject to review and approval by the senior officers of the military service, who are often the advocates for the system under acquisition (Fox et al., 1994):

> After three decades of attempts by the Office of the Secretary of Defense to obtain realistic assessments of program status and unbiased cost estimates to complete, it is clear that delays in obtaining candid assessments will continue until the conflict in program manager roles and responsibilities is resolved. ... The past three decades have demonstrated that the military services are unable to address this problem because program sponsorship and program advocacy places them in a position that often conflicts with reporting realistic program status.

Accurate reporting on the status of programs when there are serious problems can have an adverse effect on OSD and congressional support. This pushes the acquisition manager to become a program advocate if he or she wishes to progress within his or her service. It is difficult for a program manager to contradict his or her own service when the senior officers who evaluate his or her performance and who control his or her future have underestimated and overpromised the program from the beginning. This creates an inherently unstable positive feedback process (Fox et al., 1994):

> It is very difficult to expect a program manager to blow the whistle on his own service when the senior officers who evaluate his performance and who control his future have underestimated and over-promised the program from the beginning. Too often his course of action is

to go along with the game, hoping that he will be transferred before the true costs become known. This is the reality described by many government managers as well as by managers in defense contractor organizations.

Acquisition managers are pushed toward optimistic estimates in order to avoid budget/schedule slips. When the budgetary office "buys into" a program with overoptimistic estimates, the program is set up for larger and more visible failure. When actual status and realistic estimates eventually come to light, the program is at risk of being cancelled or dramatically restructured with further cost and delay impacts on the program. External forces exacerbate funding instability. Funds are shifted from one program to another and from the current budget cycle to some point in the future, in order to make some programs "well" at the expense of other programs, while remaining with the President's top-line budget. Funding shortages, whether they result from poor phasing or limited scope, impose replanning and restructuring, which inevitably increases cost and delay (Finley, 2007).

Effective implementation and integration of SE requires an organizational structure and culture that promotes ownership and accountability. In recent years, there has been a tendency toward a more cooperative relationship between the government and contractor. Integrated product teams (IPTs) have replaced objective, independent SE and the accompanying analysis. The IPT was intended to help integrate SE into the design process by ensuring early and continuous technical coordination. But the lack of independent SE and SE analysis means that there is no independent check on the quality and completeness of the SE and engineering design artifacts. The members of the IPT have already endorsed their designs and are not in a position for critical evaluation of the product and its integration into the system. Assumptions and designs are not challenged and go forward without independent assessment (Kadish et al., 2006):

> Effective oversight has been diluted in a system where the quantity of reviews has replaced quality, and the tortuous review processes have obliterated clean lines of responsibility, authority, and accountability. The oversight process allows staffs to assume de-facto program authority, stop progress and increase program scope.

When no single person or group has ownership of SE, essential SE functions are dispersed among other project/program elements without coordination and control. This can lead to a lack of well-defined SE objectives and inadvertently promote attitudes that consider SE a non-value-added activity. To avoid or correct this problem, the SE function should be well defined, have clear objectives, and have a functional mission in the project/program and specific home within the overall organizational structure (Kuykendall, 2006).

Another implementation issue that has been identified is the situation where a project has a contractual requirement for SE, but the contractor assigns a designated SE representative without a valid SE infrastructure and top-level support from management (Kuykendall, 2006). This leads to limited SE functions that are external to other project processes. With such a lack of integration, this leads to a limited impact of SE. A similar situation occurs when an organization includes SE, but does not include the SE function in the engineering processes. In this case, SE is often limited to the role of document control and requirements management, and in this role, SE still operates external to the primary project engineering development. If SE is implemented in this way, then the SE function is limited to management of project requirements, with no interactive involvement in the engineering process. When SE has no interactive involvement in the engineering process, there is a risk of failing to recognize the effects of changing requirements and requirements creep on cost, schedule, and engineering rework due to conflicts with completed activities. Horizontal and vertical integration of SE functions across the organization, including both engineering and engineering management structures, can ameliorate these problems. This enables SE to provide proactive and dynamic requirements management. In short, SE should not be restricted to requirements management, but should be integrated in all phases of the engineering development process (Kuykendall, 2006).

Although we argue above that SE needs an organizational home and clear mission and authority, this can lead to a situation where an SE organization is home to the SE specialist who has studied SE theory, tools, and techniques, but has little experience in applying and presenting SE solutions that are relevant to the project. This can be due to the fact that often, the SE does not have the requisite subject matter expertise at integration level (Schaeffer, 2005). To avoid or correct this problem, the SE specialists must be responsible for practical, that is, efficient and cost-effective, implementation of the SE program. SE specialists should participate in all project baseline activities in order to understand priorities, perspectives, and the technical subject matter (Kuykendall, 2006).

A final implementation issue to consider here is the situation where an SE program plan calls for resources disproportionate to the overall project and cannot be supported. By focusing internally on resource allocation, the SE function misses opportunities to implement SE principles through coordination and negotiation with the rest of the organization. To avoid or correct this problem, the SE team needs to define a program that is the correct size for the level of effort (or allotted budget and schedule) and coordinate SE activities with other project elements to leverage their participation (Kuykendall, 2006).

James Finley, Deputy Undersecretary of Defense for Acquisition and Technology, summarized implementation issues in a 2007 speech to the National Defense Industrial Association: "Programs usually fail because we don't start them right." This section has highlighted several topics that he cited, including inadequate early technical planning, inadequate funding or phasing of funding, lack of schedule realism, and lack of a credible back-up plan (among others) as threats to program success (Finley, 2007).

2.2 Execution Issues in Systems Engineering

Possibly the biggest single execution issue in the field of SE is the specification and management of requirements. Requirements are generated by the service advocate for the program. They are outside the SE process, and out of control of the SE process. The program acquisition manager is not in a position to gainsay a two-star flag officer who says, "These are the requirements," and especially not when the promotion/assignments of the acquisition managers are in the hands of the services. A large part of SE is concerned with managing the requirements—deriving subsystem and component requirements, assessing designs and interactions relative to requirements, testing to requirements, etc. Volatile, incomplete, bloated, hidden, over-specified, un-prioritized, and inconsistent requirements undermine effective SE. System development can proceed with incomplete requirements, that is, aspects not addressed. It is often not until late in the development cycle, sometimes during prototype testing, that the system is recognized as deficient in important, but neglected, areas due to incomplete or undefined requirements (GAO, 1971). Prioritized requirements are the exception rather than the rule. When a program is initially formulated, it may have an infeasible set of conflicting priorities. This leads to extensive work and rework to establish that the requirements cannot all be met simultaneously, and guesswork as to actual priorities. Clarification of priorities among requirements during program execution is another source of requirements volatility (Kadish et al., 2006).

> … the "top five" areas affecting the Acquisition System are *requirements management*, budget and funding instability, technological maturity, organization, responsibility, authority and accountability, and regulation and policy interpretation. … When respondents were asked to identify why Department of Defense acquisition programs have significant cost growth and schedule extensions, *requirements instability was the most mentioned problem area*, followed by funding instability and high-risk systems. Of the respondents, 96 percent agreed that program stability and predictability—to include *requirements stability*, funding stability and technological maturity—are crucial to maintaining cost, schedule and performance (Emphasis added).

There is a cultural tension between the DoD acquisition community who want to provide a complete solution with the first delivered article of materiel and combatant commanders who often have a pressing operational need. Requirement developers mandate systems that can be technologically unrealistic (GAO, 1971) or unable to be delivered within the combatant commanders' "time-to-need." Even after initial requirements are specified, the escalation of requirements often drives costs beyond the original budget and procurement timeline. This can be exacerbated by a decision to proceed, which occurs in stovepipes (Schaeffer, 2005) made on the basis of inadequate data regarding technical readiness level of component and system technology (GAO, 1971) and the stability and understanding of requirements (Schaeffer, 2005). It has been observed that the integrated consideration of requirements, technical readiness, risk mitigation plans, schedules, and budget can often occur as late as the critical design review stage (Kadish et al., 2006).

2.3 Case Example for Requirements Failure: Requirements Creep and the Aquila Remotely Piloted Vehicle

During the Vietnam War, U.S. remotely piloted vehicles (RPVs) flew over 3000 reconnaissance missions. After the end of the Vietnam war, the Air Force and Navy discontinued their use for operational missions, and by 1981, the Pentagon had no operational RPVs. The Army determined it had a need for RPVs for local reconnaissance, surveillance, and target acquisition. In 1974, the MQM-105 Aquila program was born, which became an attempt by the U.S. Army to field a *multi-mission* battlefield RPV.

The Army began the Aquila program in 1974 to provide an intrinsic local aerial reconnaissance and surveillance capability. Throughout its development phase, the Aquila experienced ever-changing and increasing mission requirements, including all-weather, day–night target acquisition, identification, tracking and laser designation, passing control of the RPV in flight from one ground station to another, jam-resistant communications, and fully automatic net retrieval. In addition to creeping payload and control requirements, range, speed, altitude, and endurance requirements were also increased. As the payload grew, it became a more costly asset, which made recovery more important. It also required a larger airframe and engine, which made recovery more difficult.

The GAO reported that after 14 years of development, the Aquila proved hard to launch, regularly failed to detect its targets, and successfully completed only 7 out of 105 flights. The Pentagon canceled the program in 1988 after 14 years in development. Development costs had soared from $123 million in 1978 to nearly $1 billion by 1987. The unit price rose from $100,000 to $1.8 million.

In contrast, the Israeli Defense Force (IDF) began the development of the Mastiff RPV at about the same time as the Aquila program. Israel's Tadiran Company took 5 years and $500,000 to design and produce the Mastiff—essentially model planes equipped with a Sony television camera and wireless video and data link. The IDF used the Mastiff effectively in the 1982 invasion of Lebanon to pinpoint Syrian missile sites for artillery attack. In the mid-1980s, the Navy purchased Mastiff-II RPVs for approximately $100,000 each.

2.4 Case Example in Requirements Failure: "Stated-but-Not-Real" Requirements, "Real-but-Not-Stated" Requirements, and the Future Combat System

The FCS was an Army program to equip the modular brigade combat teams (BCTs) with a family of eight manned ground vehicles (MGVs) based on a common chassis. These vehicles, together with unmanned ground and air vehicles and unattended sensors and munitions, were to be linked together and integrated with theater and joint assets via an advanced communications and information network. The concept was to exploit U.S. information and network superiority together with strategic and intra-theater mobility of lightweight vehicles, to achieve radical increase in combat tempo, precision, and effectiveness without sacrificing survivability.

Between 1996 and 1999, the Army formulated the theoretical foundation for networked operations though studies, analyses, and war games for the "Army After Next". In 1999, the Army found that operations in the mountainous regions of the Kosovo deployment were impeded by the limitation on strategic and theater mobility of the heavy vehicles of the current fleet, for example, the 70-ton Abrams Main Battle Tank and the 45-ton Bradley Infantry Fighting Vehicle. This led to calls for a more deployable, lighter force. From 1999 to 2003, the Army developed operational plans and material requirements for the FCS. In 2002, the lead systems integrator was selected and the program official began in 2003. The initial total program cost to outfit 15 BCTs in 91 months was $80B. In 2004, the program went through a major restructuring that delayed completion by 48 months and increased the cost estimate by 35% to $108B. At this time, only one of the program's 54 critical technologies was considered technically mature. The revised program cancelled three of the four unmanned ground vehicles (UGV). The program was scheduled for a concept design review at the end of 2009. In the spring of 2009, the Secretary of Defense issued an administrative memorandum terminating the MGV portion of the program and reinstating the UGVs.

The MGV portion of the FCS program provides examples of some of the "contextual" factors that lead to cost and schedule growth and impede effective SE. The transportation and deployment concept was that the MGVs could be transported by the C-130 "Hercules" cargo aircraft and roll-off ready to fight. This imposed a weight limit of 19 tons, and constrained the vehicle geometry. When the designers could not meet the weight limit and still provide all the components and capabilities the Army wanted, the Army relaxed the weight limit by a few tons, then a few tons more, and so on until the heavier MGVs were over 30 tons. With each change to the requirement, there was a schedule slip. For whatever reason, the weight limit was not, in effect, a real requirement, and the attempt to meet the sliding requirement increased the engineering cost and schedule.

For the mounted gun system, the contractor conducted a trade study of alternative armament that concluded that a version of the 105 mm cannon was the best option; it was more accurate than the 120 mm cannon, was lighter, had lighter ammunition, and was almost as lethal as the 120 mm cannon. The theory was that in network-centric operations, other assets, for example, aircraft, could take out the few hard targets that the mounted gun system could not. However, the Army insisted on a 120 mm cannon. This was an example of a real-but-unstated requirement.

2.5 Complex "Systems-of-Systems" Issues

According to the *Defense Acquisition Guidebook*, a "system of systems" is an interacting group of autonomous but interdependent systems with qualitatively new and different capabilities as a whole different from and greater than the sum of the component parts. It can include a mix of existing, partially developed, and yet-to-be developed independent systems. SE challenges include engineering under conditions of uncertainty, continuing architectural reconfiguration, and emergent properties. Emergent properties are, by definition, not predictable from the component parts. They are surprises. We need to develop architectures, concepts, and engineering development strategies that minimize surprises. It is even harder to *design* a complex system to have desired emergent properties and to avoid undesirable emergent properties.

There has been an increasing push toward systems that are designed to use common subsystems or that are interoperable with other systems, as part of the attempt to reduce logistics burdens and create network-centric systems-of-systems. This creates dependencies between separate programs, so that cancellation or delay of one program directly affects other programs. In some cases, two or more separate programs are supposed to interface, but neither is sufficiently well defined for an interface specification or interface control document. A related problem in concurrent development is designing under wishful thinking—that a technology will mature and a real capability years down the road just in time for production. There is a particular risk to this approach in the case of convergent programs. Convergent programs are characterized by the situation in which the capability or success of program A is dependent on the success of program B, but the two programs are independent line items. Program B can be cancelled, delayed, or be found technologically infeasible. Initially, program A can deny any risk associated with technology B (since it is a separate program), but when program B is re-scoped, delayed, or cancelled, program A is put in jeopardy. There are ripple effects when the program is cancelled or the technology does not mature.

SE for highly complex systems is an area of much speculation and little experience. It is not clear that DoD acquisition actually wants or can handle complex systems. The central characteristic of a complex system is that its properties and behaviors emerge from interactions among the subsystems, but cannot be predicted or explained by simpler subsystem interaction models. If the intent is for a system to meet its requirements as a result of emergent behaviors, and for the emergent behaviors not to violate requirements, then traditional "march-of-progress" methods cannot succeed. It is not possible to design a system to have specific emergent properties and behaviors using traditional requirements management, SE, and engineering management methods.

Traditional SE assumes that analysis of the interactions at the component level yields a simple explanation of the apparent complexity of phenomena observed at the system level. This is the basis for top-down analysis and design, for deriving subsystem and component requirements, and for evaluating the effects of subsystems and components on the overall system. Traditional top-down models attempt to organize a system into subsystems that minimizes the information flow, and especially feedback loops, between subsystems. In this way, the architecture reduces the entropy of the total system. The description of the system in terms of subsystems and interactions produced a more compact, simpler, and still correct description of the total system performance. The traditional SE paradigm fails for such highly complex systems. This poses significant challenges for requirements derivation and performance assessment. Software developers gave up top-down design a quarter century ago and in place substituted object-oriented design. The equivalent SE label is "entity-based design." SE methods, practices, and tools for entity-based design have yet to be developed and tested in real system development applications.

Given the gaps in traditional SE, evolutionary development is promoted as the preferred acquisition practice for highly complex systems, with spiral development and incremental development being the two principal approaches. Evolutionary development creates the hazard that program managers will opt to achieve temporary cost and schedule success by deferring technical risks and difficult requirements to the next increment or cycle. Even more troubling is the risk of institutionalizing requirements volatility and technology optimism. Spiral development, originally formulated for large software projects, has been adapted for integrated hardware and software systems. Prior to 2008, spiral development was the preferred approach to EA. In 2008, the DoD abandoned spiral development in favor of incremental development for future programs such as the ground combat vehicle. EA is now represented as one process. The objective remains to put technology into the hands of users quickly, but with a strong emphasis on defining militarily useful increments of operational capability, that can be developed, produced, deployed, and sustained. This is a collaborative process among the user, tester, and developer, where each increment will have its own set of threshold and objective values set by the user, dependent upon the availability of mature technology (DoD Instruction Number 5000.02, 2008).

The distinguishing feature of highly complex, adaptive systems is that the overall behavior and properties cannot be described or predicted from the union of their subsystems and subsystems' interaction. The system–subsystem architecture and subsystem interactions either do not simplify the complexity of the description or fail to represent the system properties.

The reduction in total system entropy comes about through the dynamic emergent properties and behaviors, rather than a static architecture and encapsulated subsystem interactions.

Complex systems unite their subsystems (i.e., add organization) in a way that reduces total entropy, and defy reduction in entropy through hierarchical decomposition. The reduction in entropy is the result of the emergent properties and behaviors. There are significant barriers to relating the components to the overall system (Abbott, 2007): (1) The complexity of the system's description is not reduced by formulating it in terms of the descriptions of the system's components, and (2) combining descriptions of the system's components does not produce an adequate description of the overall system, that is, such a description is no simpler than propagating the descriptions of the components.

Natural systems that exhibit such complexity, such as organisms, are characterized by robustness and adaptability, in contrast to human-designed systems that exhibit fragile emergent properties. This has been attributed to the top-down nature of the requirements-driven human-built system development process, in contrast with the bottom-up processes that occur through natural selection (Abbott, 2007):

> One reason is that human engineered systems are typically built using a top-down design methodology. (Software developers gave up top-down design a quarter century ago. In its place we substituted object-oriented design—the software equivalent of entity-based design.) We tend to think of our system design hierarchically. At each level and for each component we conceptualize our designs in terms of the functionality that we would like the component to provide. Then we build something that achieves that result by hooking up subcomponents with the needed properties. Then we do the same thing with subcomponents... This is quite different from how nature designs systems. When nature builds a system, existing components (or somewhat random variants of existing components) are put together with no performance or functionality goal in mind. The resulting system either survives [or reproduces] in its environment or it fails to survive.

It may be possible to use artificial intelligence genetic programming methods, that is, artificial evolution, to design systems with desired emergent properties. The most notable success of genetic programming and artificial evolution in engineering has been in application to the design of analog electrical circuits to perform mathematical operations, for example, to take cube roots. These methods have invented robust and effective, but hither-to unknown circuits to perform desired operations. However, this was a fairly constrained application relative to weapon systems. Use of the methods requires truly accurate simulation capabilities (to evaluate the "fitness" of each design), a suitable genome model, an algorithm for expressing the phenotype from the genotype, and a characterization of the variations in the application environment. It would be a significant challenge to apply these methods to DoD systems.

2.6 Conclusion

This chapter has presented evidence of the consensus view that the primary contributors to large and unpredictable growth in cost and schedule of DoD acquisition programs are caused at least in part by shortcomings of both the implementation and execution of SE, including (1) overoptimistic cost, schedule, and technical maturity estimates; (2) volatile, incomplete, bloated, hidden, over-specified, un-prioritized, and inconsistent requirements; and (3) dual lines of responsibility that suborn independent program management, cost, schedule, and maturity/risk assessment. In addition to these issues, other challenges are rapidly emerging to effective implementation and execution of SE, related specifically to complex, adaptive SoS. These include spiral development with coevolution of requirements and design, and prediction and management of emergent system properties and behaviors. These challenges are being met by a renewed emphasis on and revitalization of the SE function. As stated by top-level DoD officials (Wynne, 2006):

> We've made the revitalization of systems engineering a priority within the U.S. Department of Defense. … We expect to see a reduction in acquisition risk, which ultimately translates to improved product cost control over the entire life cycle. … We will accomplish this through systemic, effective use of systems engineering as a key acquisition management planning and oversight tool. In addition, we will promote systems engineering training and best practices among our acquisition professionals.

This commitment, combined with a growing interest in the research community in the development of a next generation of SE methods, practices, and tools, can forge the way ahead for the next generation of systems engineers.

Authors

Gary Witus has over 30 years of experience in defense systems acquisition and engineering. He was an active participant in several major defense acquisition programs including the Advanced Field Artillery Tactical Data System, the Tacit Rainbow Cruise Missile System, the M1A2 Abrams Main Battle Tank, the Target Acquisition Model Improvement Program, and the Marine Corps Expeditionary Fighting Vehicle. Dr. Witus received his BS in mathematics from the University of Michigan in 1975, his MS in industrial and operations engineering from the University of Michigan in 1978, and his PhD in industrial and manufacturing engineering from Wayne State University in 2002. After receiving his MS degree, he spent 3 years working with General William DePuy (Ret), a four-star general, former vice-chief of staff for the Army, and first commander of the Army Training and Doctrine Command, on the use of models, simulations, and operations analysis in defense acquisition programs.

R. Darin Ellis has over 15 years of experience in human systems integration and systems engineering of relevance to DoD. He received his BS from Kettering University, Flint, Michigan, and MS and PhD both from Penn State, University Park, all in industrial engineering. His research and teaching interests include human-in-the-loop systems modeling and simulation and related test and evaluation methods. His sponsors have included DoD, Veteran's Administration, and NASA.

References

Abbott, R. 2007. Putting complex systems to work. In Introductory paper at the *2007 Symposium on Complex Systems Engineering*, Santa Monica, CA.

Defense Acquisition Guidebook. https://akss.dau.mil/dag/ (accessed September 18, 2010).

Department of Defense Instruction 5000.02. December 8, 2008. *Putting Operation of the Defense Acquisition System*. http://www.dtic.mil/whs/directives/corres/pdf/500002p.pdf (accessed November 17, 2009).

Finley, J.I. 2007. Keynote presentation. In *10th NDIA Annual Conference on Systems Engineering*, San Diego, CA.

Fox, J.R., Hirsch, E., Krikorian, G., and Schumacher, M. 1994. *Critical Issues in Defense Acquisition Culture—Government and Industry Views from the Trenches*. Defense Systems Management College, Executive Institute, Fort Belvoir, VA.

Government Accountability Office (GAO). 1971. *Acquisition of Major Weapon Systems*. GAO Report B-163058, Washington, DC.

House Armed Services Committee. May 2005. *House Committee Report 109-89, FY2006 Defense Authorization Bill HR 1815, Title VIII, Acquisition Policy*, Washington, DC.

Kadish, R., Abbott, G., Cappuccio, F., Hawley, R., Kern, P., and Kozlowski, D. 2006. *Defense Acquisition Performance Assessment Project Report for the Deputy Secretary of Defense*. Office of the Deputy Secretary of Defense, Washington, DC.

Kuykendall, T. 2006. Why Systems Engineering Fails. In *2006 INCOSE Annual Meeting*, Orlando, FL.

Schaeffer, M.S. July 2005. Technical Excellence through Systems Engineering. *Boeing Technical Excellence Conference*, Long Beach, CA.

Secretary of Defense, Gordon England. June 2005. *Acquisition Plan*. Memorandum initiating the Defense Acquisition Performance Assessment Project.

Wynne, M.W. and Schaeffer, M.D. March–April 2005. Revitalization of systems engineering in DoD—Implications to product cost control. *Defense AT&L*, 14–17.

Chapter 3

Functional Analysis and Architecture

Ricardo L. Pineda and Eric D. Smith

Contents

Derived from the user need and customer requirements, logical functionality defines a system's scope and outputs within its operating environment. A function within the context of systems engineering (SE) denotes an action that the system must perform in order to fulfill its mission or objective.

A full understanding of functions is required for a complete description of the system in the form of a functional architecture. Functional architectures contain logical decompositions of high-level functions into lower-level functions. High-level functions occur in the operations environment, which dictates how the system must work at the level of operations personnel. Lower-level functions are allocated to the physical architecture of the system, which describes the constituents of hardware and software. As functions are decomposed and allocated, functional analysis looks into the required data and message flows and the physical components that will satisfy the customer requirements. High-level principles of architecting are important at this stage of functional analysis to ensure architectural robustness, ensuring, for example, that high-level invariant functions endure as new technologies are developed and the physical architectural elements change.

This chapter reviews principles of functional analysis, including functional decomposition and functional allocations to physical components, as design activities that establish a system's expected behavior in its particular operating environment. Case studies of the wooden seventeenth-century Swedish warship, the Vasa, and information and communication technologies (ICT) are examined from the services to be provided, the fundamental view of their functional/logical architecture, and finally their physical architecture. A classical ICT network is presented and analyzed against a current state-of-the art Internet Protocol (IP) multimedia services (IMS) network to illustrate architectural robustness in the fast-changing telecommunication field.

3.1 Introduction

Systems must be modeled in order to reduce their complexity. Abstract models, where only essential elements are included, allow an increase in the efficiency of the design process and in activities following design, such as development, testing, deployment, operations, maintenance, and upgrading. From a traditional systems engineering (SE) point of view, it is often assumed that a system can be thought of as a hierarchical structure with levels becoming more detailed as the model progresses downward toward properties that can be executed independently, or close to independently, by a lower-level subsystem, node, or component. It should be noted that this basic working principle is linear (Newtonian) and it is assumed that properties can always be delegated to lower levels in the hierarchy. This may not be the case as system complexity increases, for example, in systems of systems (SoS) or complex SoS.

This chapter begins with a review of the SE process and then describes modeling and architecting in SE, under the following tripartite division:

1. Operations/behavior modeling and analysis
2. Functional/logical modeling and analysis
3. Physical architecting

Modeling is defined as the generation of an abstract, conceptual representation of a system, which can be used to determine the functions of the system. Architecting can be defined as the artful allocation of system functions to physical subsystems or components, which are arranged not only with regard to their physical characteristics but also with regard to how important subsystem attributes and emergent system attributes contribute to customer satisfaction. Good modeling and architecting requires that attention be paid to major architectural features of the system and model, including purpose of the system, assumptions made, system behavior, levels modeled, subsystem architecture, interface architecture, and whole system architecture.

For simplicity, it is largely assumed in this chapter that the architect will be able to zero in on the best architecture. In reality, a variety of viable architectures will arise as part of a thorough architecting effort. These alternative architectures may even employ widely different models as a starting point. Alternative viable architectures should be fairly considered as candidates for the final architecture in an iterative process that involves collection and validation of underlying models, gathering empirical or experimental data for each model, simulation or experimentation, and analysis of performance. Once the architectures have been developed to their fullest potential via the optimization of underlying models, a criteria-based trade-off study among the alternative architectures can select the preferred architecture. During the optimization and trade-off processes, the robustness of the models and architectures should be tested with sensitivity and risk analyses. For now, extensive steps are ignored in order to focus on the fundamental steps of operational/behavioral modeling and analysis, functional/logical modeling and analysis, and physical architecting.

3.2 Systems Engineering Process

The centrality of functional analysis in SE is visible in the most accepted representation of the SE process, as shown in Figure 3.1.

This SE process diagram is characterized by a succinct progression through fundamentally different aspects of system design, with feedback and iteration loops

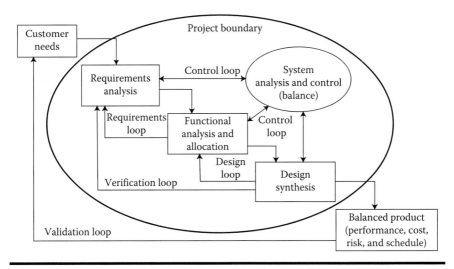

Figure 3.1 Systems engineering process.

included as absolutely essential features. The determination of customer needs occurs in the customer's environment, before customer needs are fed into a soft project boundary. Requirements are then formally determined, written, and analyzed, to determine the essential functions that must be performed. Note that a requirements loop remains active between the process steps of requirements analysis and functional analysis and allocation. Functional analysis involves the formal determination and decomposition of functions, and functional allocation involves the assignment of function performance to an identified subsystem or architectural component. Once system functions have been fully decomposed and identified, design activities proceed to ensure that all functions are implemented to the full extent of the system requirements. A design loop assures a repetitive examination of design activities that progress toward a proper design synthesis. Design parameters and characteristics are not fully vetted until tested against the mandated requirements via a verification loop.

The final balanced product, with a satisfactory balance among its performance, cost, risk, and schedule characteristics, is then presented to the customer as part of the validation loop, which ensures that the system as a whole validly satisfies customer expectations. Note that all SE process activities are overseen and controlled by Systems Analysis and Control activities, with active control loops. The SE process is depicted as linear because of the necessities of textbook presentation, but it is essential to unequivocally state that the SE process is actually highly nonlinear and iterative. In addition, the SE process is most essentially holistic, in that it seeks a balanced product that is coherent as concerns all the relations among product elements.

3.2.1 Product and Processes

A system is distinct from the processes that enable and support it throughout its life cycle. Processes that create a product, for example, must be described independently from the product, lest intermingling of their two separate natures occur. A product exists in time and space, with a preliminary existence as a conceptual system and system design. A product may have dynamic functions and a logical design, but these should not confuse the fact that a product is distinctly different from the process that designs the product—via the orderly actions of disciplined engineering activities. Other processes may include the production processes of a disciplined workforce, and the operations processes that implement the functionality of a system in the environment. This chapter distinguishes between product descriptions and process descriptions.

3.2.2 Requirements Analysis

For brevity, this chapter does not include descriptions of critical preliminary processes for determining customer needs. Requirements determination and analysis is only introduced at a basic level, since most of the discussion will begin with functional analysis and allocation. This chapter does, however, describe requirements analysis, and all subsequent system design, in the broad scope and form of the tripartite separation of operational/behavioral modeling and analysis, functional/logical modeling and analysis, and physical architecting, which together lead to a singular design synthesis. Note that operations and functions both refer to actions, but at different levels, namely, system level functions, and higher-level operational activities and behaviors. Figure 3.2 illustrates this tripartite separation. In the case studies, retrospective examinations of designs are conducted with regard to functional analysis and architecture as expanded in this fashion.

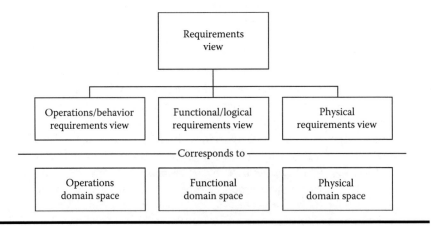

Figure 3.2 Requirements view and tripartite division of design areas.

Operations at a high level are necessarily accompanied with an examination of the large-scale behavior of the system, that is, what "use" the system is intended to have. Similarly, functions at the systems level are necessarily accompanied with the logic that determines the order in which functions are performed. Functional/logical architecture is always central to systems, implementing "what" the system must do.

3.2.3 Design Space

In general, then, we espouse a three-dimensional view of the general design space as seen in Figure 3.3.

Figure 3.4 shows that three requirement views, each associated with a design space dimension, can be decomposed down to whatever level is necessary in order to design the system.

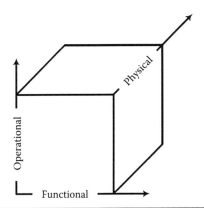

Figure 3.3   Three-dimensional view of the design space.

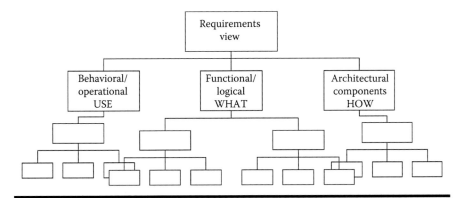

Figure 3.4   Requirements and design space decomposition.

3.2.4 Scope: Operations versus Operating Environment

Scope defines the boundary between the system and its environment. For example, operations refer to the actions that will have to be taken on an ongoing basis by the system or the system operators in order to successfully operate the system in order to achieve its intended purpose and also to sustain the system's effectiveness. As such, operations refer to the processes that the system must execute or the disciplined activities (methods and procedures, M&P) that the system personnel must follow in order to deliver the expected goal. "Operating Environment" refers to the externally imposed conditions within which the system will have to operate. An operating environment may be a military theater of operations, HVAC (heating, ventilation, and air-conditioning) conditions in an operating center, or an imposed computer operating system standard such as Windows or Linux. We emphasize these concepts here because there is a tendency to mix these terms in the literature and by practitioners.

3.3 Modeling Elements and Foci

This section will introduce different dimensions and analyses that help to categorize system parts, for the sake of reducing system complexity. Note that some dimensions and categorizations have already been introduced, namely, the distinction between product and process, requirement types and associated design space dimensions, and the scope-based distinction between operations and operating environments.

3.3.1 Requirements: Functional and Nonfunctional

Requirements are necessarily derived from the customer's need. Some requirements, collectively called functional requirements, fall into the category of actions that need to be performed, either at the global operational level (operational requirements), or at the system functional level (functional requirements). Nonfunctional requirements are specifications that stipulate levels of performance, or a measure of effectiveness (MoE) value that must be achieved, or a particular attribute the system must possess. Functional requirements can also be assigned to time periods depending on when they need to be performed, as shown in Figure 3.5 as pre-mission, mission, or post-mission requirements.

3.3.2 Levels and Hierarchy

Functional analysis and architecting invariably involves the identification of levels within systems, amid other SE design activities. Note that system levels are abstractions useful for design clarity and functional simplicity. For an exposition on the

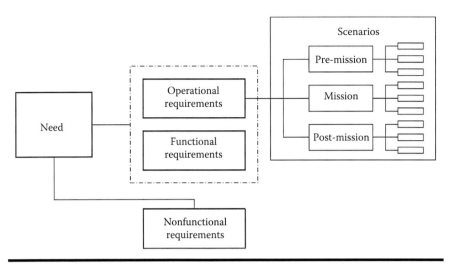

Figure 3.5   Requirements: functional and nonfunctional.

requirement that valid systems must be modeled with defined levels, see Bahill et al. (2008). The biologically inspired work of the founder of general systems theory, Ludwig von Bertalanffy (1933, 1968, 1981) formalized the concept that, at higher system levels, emergent properties become present, while at lower levels, the differentiation of elements is increased. For example, the aliveness of a physical organism cannot be explained by examining its lower-level elements in isolation, just as the battlefield tactics emerge from the direction of individual field elements by specific orders.

Systems levels are distinct strata within which elements interact in a common way, but outside of which elements interact differently. For example, bacteria interact with each other through conjugation, but interact differently with host bodies at a level above, or chemicals at a level below. In a well-stratified and orderly system, it is uncommon for elements to interact with any elements that are not in the same level. For example, Internet routers interact with each other, but not with applications on user computers. Of course, elements often do interact with elements at levels immediately above or below them. The most common mistake in modeling is to connect elements separated by two or more levels (Bahill et al., 2008). Figure 3.6 shows the decomposition of a generic system into a hierarchy of levels, across which interactions may occur.

3.3.3 Time

Time, typically described in the form of a system-specific life cycle, is a crucial dimension, because often system operations and behavior are different at different points in the system life cycle. Changes may also occur in the system functions and in the logic connecting the functions. Additionally, if the technology employed

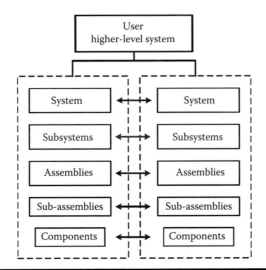

Figure 3.6 **Hierarchical abstraction of interacting systems levels.**

changes, or if the system configuration changes, the physical components of the system may be markedly different in different life cycle phases. For instance, a computer's components may be upgraded as technology advances, but the components may still perform the same functions.

3.3.4 Matrices of Relations

With the dimensions described above, very useful matrices can be constructed. For example, consider the matrix shown below in Table 3.1. It is clearly important in identifying at what stages in the life cycle certain functions will be performed.

Function versus architectural component tables are another example of a matrix type that sheds light on relations within complex systems. As we will see later, it is

Table 3.1   Life Cycle-Dependent Functional Performance

	Time		
	Pre-Mission	*Mission*	*Post-Mission*
Function A	X		
Function B		X	X
Function C	X		
Function D		X	

Table 3.2 Functional Allocation Matrix

	Component 1	*Component 2*	*Component 3*
Function A	X		
Function B		X	
Function C			X

beneficial for each component to perform only one function, and for each function to be allocated to only one component, as in Table 3.2.

3.3.5 Prioritization

Prioritization is a general concept and procedure that aids in handling complex problems when there are limited resources available. Often, budget and schedule constraints force work to be prioritized. If this is the case, high-priority needs, requirements, or functions should be designed and assured before the addition of lower-priority items. High-risk items are the most obvious high-priority items, because the failure of a high-risk item can cause the entire system to fail. Needs, requirements, or functions should be designed robustly if the physical components that implement these are likely to change. Robust designs mitigate the negative effect changes in technology items. For example, the abstract functional understanding of recording and playing music has not changed much over the decades, even as the technological implementations have changed greatly.

3.3.6 States

States—as distinct from requirements, functions, and architectural components—provide a fundamentally different modeling element, and an alternative orthogonal dimension for modeling. States are present conditions that imply ongoing actions, and are highly important for functional analysis. For example, a washing machine may be *washing, waiting, or idle.* It is important in SE to precisely state when different states occur.

3.4 Functional Analysis Principles

3.4.1 Functions as Central to Design

Stafford Beer (1985) coined the acronym "POSIWID" or the "Purpose of the system is what it does." This aphorism reminds us that the essence of a system is its purposeful, value-adding action. If a system cannot perform the functions desired by the customer, it is not providing value. As a caveat, the purpose of a system is often not its touted purpose, which is often driven by extraneous human reasons, such as marketing or political motives.

3.4.2 Conceptual Design

For unprecedented systems, a designer must begin by examining the customer's needs, and asking: "How can the customer's needs be met by an abstract functional system?" This questioning brings the conception of the necessary actions that the system must take. Some actions will be conceived as global operations, while other actions will be local functions. All functions should be arranged by functional decomposition, described in the next section, or by functional flow diagrams (FFDs).

3.4.3 Context Diagrams

Context diagrams have a function at their center, and four contextual elements surrounding the function. "Context diagram" is the term currently used by the International Council on Systems Engineering (INCOSE, 2006), although the same sort of diagram was previously labeled: IDEF0 (Integration DEFinition language type 0 diagram), ICOM (input, output, control, mechanism) diagram, or, at the enterprise level, an enterprise architecture capability diagram (EACD). Figure 3.7 shows the function of a context diagram, here called an activity unit, surrounded by orientation-specified inputs, outputs, controls, and enablers. Note that an activity unit can encompass and subsume a complete process.

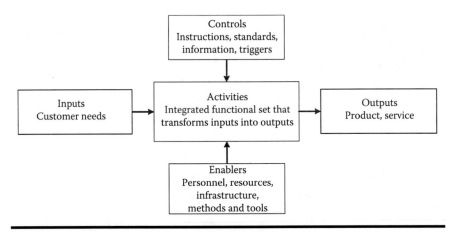

Figure 3.7 Activity/function in a context diagram.

3.5 Functional and Logical Analysis Process

This section describes functional analysis in terms of hierarchical functional decomposition, process-oriented FFDs, functional interface analysis via N² diagrams, and functional timeline analysis. Figure 3.8 summarizes the functional analysis and architecting process.

3.5.1 Functional Decomposition

Functional decomposition is basic to SE, and has been emphasized by the Department of Defense in such documents as Military Standard 499A (1993). Functional decomposition begins with the identification of one top-level system function that encompasses all functions performed by the system. The top-level function is usually robust, in the sense that it needs to be performed despite most changes in the environment, and many changes in the operational space and the system technology. The top-level function is also robust in the sense that systems usually need to perform a same basic action during most of the system's life cycle.

The top-level function is usually far more abstract than the lowest-level functions into which it is hierarchically decomposed. Hierarchical arrangement of functions implies that the top-level function is implemented holistically by the entire system, while lower-level functions may be implemented by defined subsystems. Lower-level functions need to be differentiated until they can actually be implemented in specific subsystems, assemblies, or components, whether by newly designed or adapted components, or by commercial off-the-shelf (COTS)

Functional analysis and architecting process

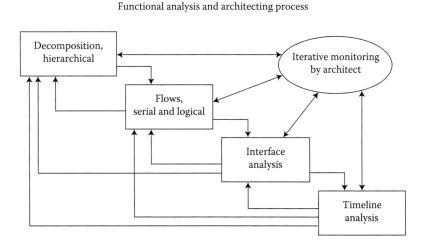

Figure 3.8 Functional analysis and architecting process.

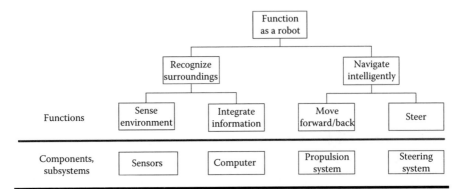

Figure 3.9 **Functional decomposition progresses until lower-level functions can be implemented physically.**

components. Figure 3.9 shows a functional decomposition, where the low-level functions are shown adjacent to elements in which the functions are likely to be implemented. For example, the function of *sense environment* will be performed by *sensors* and, the function *move forward/back* will be performed by the *propulsion system*.

3.5.1.1 Independence of Functions

The functional independence of the decomposed sub-functions is fundamental in allowing a workable functional allocation. If sub-functions are not independent, they cannot be cleanly implemented in different physical subsystems. Consider the situation in Figure 3.10, illustrative of the early twentieth century's necessity for change in the quality arena. The functions "design components" and "inspect components" are not independent.

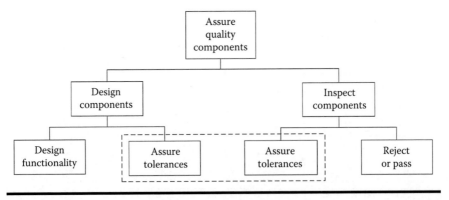

Figure 3.10 **Nonindependence of functions in an attempted functional decomposition.**

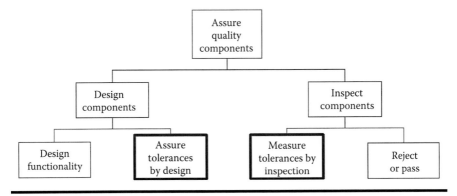

Figure 3.11   Nonindependence of sub-functions corrected by differentiation.

The undifferentiated sub-function *assure tolerances* can be differentiated into *assure tolerances by design* and *measure tolerances by inspection*, as shown in Figure 3.11. Note that a matrix of relations can be used to relate the lowest-level elements of one decomposition tree to the lowest-level elements of another decomposition tree. This type of diagram is shown in Figure 3.12, where the correlations among the lowest level elements of two decomposition trees are indicated with an X.

As a segue into functional flows, we note that the differentiation of the non-independent sub-function assure tolerances could also have been accomplished by defining functional flows. Specifically, *assure tolerances* could have been differentiated into two functions, *assure tolerances by design* and *assure tolerances by inspection*, which fit into a series.

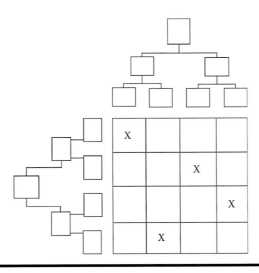

Figure 3.12   Relating decomposition trees using element relations.

3.5.2 *Functional Flow Diagrams*

A FFD will clearly indicate the order of function sequence. In the assure quality components example, a FFD could have determined the differentiation of the non-independent function assure tolerances in the way shown in Figure 3.13.

FFDs allow logic to be included as relations among functions in functional flowcharts. Functions can be performed in serial, parallel, or via a particular path. AND gates signify that functional flows proceeds along all directed arcs outgoing from the AND gate. OR gates signify that functional flows proceed along only one of the outgoing arcs. An example is shown in Figure 3.14.

FFDs indicate logical dependencies among functions, whether downstream, upstream, in parallel flow, or otherwise. For example, in Figure 3.14, if the function "build components" fails, the downstream function "measure tolerances by inspection" will fail by default. Also, if "measure tolerances by inspection" fails, the upstream function "assure tolerances by design" will be disadvantaged since it will receive no feedback from measurements. Further, note that there are two parallel input streams necessary for good performance of "measure tolerances by inspection."

An example of an FFD consisting of context diagrams with ICOM features is shown in Figure 3.15. Context diagrams can be used in a hierarchical environment, to produce generic figures such as Figure 3.16, which could be a representation of a hierarchical communications network where each node operates within a context.

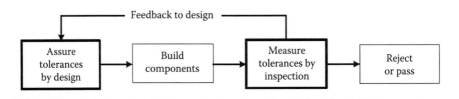

Figure 3.13 **Differentiation of nonindependent functions via a functional flow diagram.**

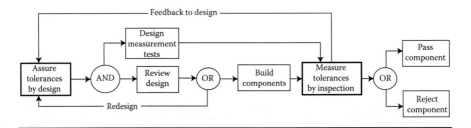

Figure 3.14 **Logic in a functional flowchart.**

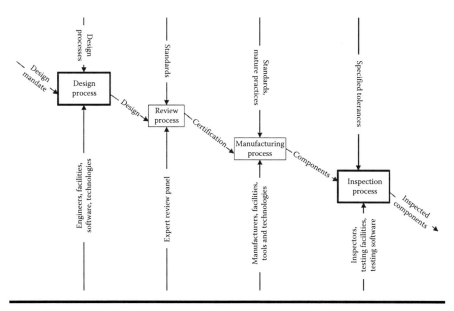

Figure 3.15 **Function flow of context diagrams in a quality improvement process.**

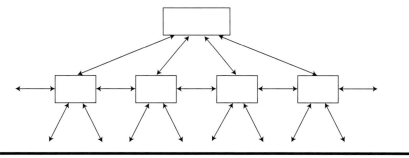

Figure 3.16 **Hierarchically arranged contexts.**

3.5.3 Functional Interface Analysis: N-Squared Diagrams

N^2 (N-Squared) diagrams can be used to show the interfaces between elements such as functions or objects. Interface analysis functions are placed on the diagonal row of a square matrix. Connections between the functions are shown as directed inputs and outputs, which constitute the interfaces. An N^2 diagram is shown in Figure 3.17. The interfaces in Figure 3.17 are read clockwise:

Function 1 output provides input to Function 2.
Function 1 output provides input to Function 4.
Function 2 output provides input to Function 4.
Function 3 output provides input to Function 2 and also to Function 1 and so on.

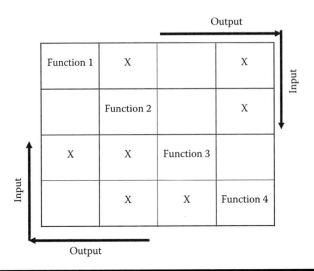

Figure 3.17   N-Square functional interface diagram.

Whenever we have an empty box, it means that there is no interface between the functions, for example, Function 1 does not directly interface to Function 3. N² diagrams help in understanding dependencies among the different functions. For example, it becomes evident which functions provide flow in the same sequence and which functions do not take part in loops. N² diagrams grow in complexity as each of the functions is decomposed into lower-level sub-functions. The lower-level functions can be analyzed using an N² diagram and again dependencies and loops may be discovered among the sub-functions.

3.5.4 Timeline Analysis

Functional flow block diagrams (FFBDs) tell us what functional sequence must be followed to achieve mission objectives; however, FFBDs do not relate functional sequences with the time durations to achieve the objectives. For instance, the functions to detect and identify a target within 3 s are the same functions needed to detect and identify the target within 10 min, but the time allocated to the functions is quite different in each case. Timeline analysis is critical in understanding the appropriate timing sequence and specific times associated with a system's functions. Time-line diagrams (TLDs) are of particular importance when looking at

1. Time allocated to different sub-functions once the total time required to accomplish the systems function is understood
2. Design requirements, for example, response time, mission duration, system lifetime

3. Operational scenarios, for example, commercial airline preflight/flight/landing
4. Operations, administration and maintenance (OA&M) times of the system

A TLD is shown in Table 3.3, where the horizontal axis represents time and the functions are listed on the vertical axis. From Table 3.3, we can see

F1 has been allocated 5 TIME UNITS and it precedes any other system function
F2 starts once F1 is completed and it has been allocated 15 TIME UNITS for execution
F3 and F5 start right after F2 has been completed; F3 has 10 TIME UNITS allocated and F5 has 5 TIME UNITS duration
F4 starts 5 TIME UNITS after F3 has started and F4 has been allocated 10 TIME UNITS for its execution

An example is shown in Figure 3.18.

3.6 Physical Architecture Process

3.6.1 Preliminary Design

Traditional preliminary design begins with the identification of viable subsystems and components, and ends with the tailored allocation of functions to subsystems and components.

3.6.2 Functional Allocation to Subsystems of Components

Functional allocation is central to both SE and to good architecting. Functional allocation is discussed in this chapter because of its traditional importance to SE, and because of its particular importance in the topic of functional analysis and architecture. However, comprehensive physical architecting must include a balanced consideration of all elements that contribute toward a well-engineered solution.

Ideally, each sub-function should be allocated to one specialized component or subsystem that will perform that function exclusively, effectively, and efficiently. There are many reasons why a 1-to-1 matching of functions and components/subsystems is preferred. First, if a function is perceived to have failed, the physical cause of failure will be obvious; contrariwise, if the physical component fails, the function will be known to have failed. Second, if a function is performed by more than one component/subsystem, the interactions among these components must be considered during design and redesign. Third, if a component/subsystem performs many functions, the efficiency and reliability of the functional performance

Table 3.3 Timeline Diagram

	Time 1–5	Time 6–10	Time 11–15	Time 16–20	Time 21–25	Time 26–30	Time 31–35
Function 1	→→→→→						
Function 2		→→→→→	→→→→→	→→→→→			
Function 3					→→→→→	→→→→→	
Function 4						→→→→→	→→→→→
Function 5					→→→→→		

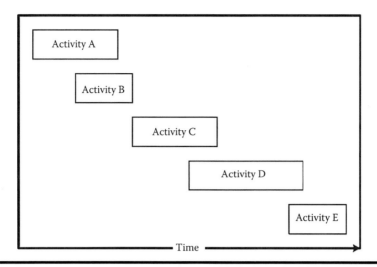

Figure 3.18   Timeline diagram.

comes into question. Table 3.4 summarizes some of the effects of allocating various numbers of functions to components with examples of functional versus physical decomposition. Table 3.5 summarizes some of the effects of allocating various numbers of functions to systems.

In general, a component/subsystem must be well designed to perform its functions. The component/subsystem must have the requisite variety—that is, more variety in controlled and intelligent responses than environmental influences—in order to properly handle the variety of inputs, controls, and mechanisms with which it is provided, and which must be converted to the desired output. Whether a component/subsystem can do this is a matter of its organization, quality, and robustness in the face of operational variability.

The B-2 Stealth Bomber provides an example of a function that is performed at an unexpected level of the physical architecture. Specifically, stealth functions are

Table 3.4   Functions Allocated to Components

	1 Component	*2 Components*	*>2 Components*
1 Function	Knife blade: Cuts	Dual jet engines: Thrusts	Firearm: Fires
2 Functions	Combination bottle/ can opener: Opens	Lever and support: Lifts or separates	Vice-grip pliers: Grips and holds
>2 Functions	Rubber band: Holds, lifts, bundles, repairs, etc.	Multipurpose bracket and screw: Lifts, hangs, holds	Digital watch: Times, calculates, synchronizes

Table 3.5 Functions Allocated to Systems

	1 System	*2 Systems*	*>2 Systems*
1 Function	IDEAL: Missile tracking system tracks airplane	BAD: Dual CEOs both direct GOOD: Redundant systems	GOOD: Three branches of U.S. government govern
2 Functions	OK: Telephone provides user interface and takes messages	GOOD: Heating and cooling unit; washer/dryer combination	GOOD: Car subsystems transport people and transport cargo
>2 Functions	OK: Humans Complex systems can adequately perform many functions	GOOD: Heating and cooling unit heat, cool, and dehumidify	OK: Bee colony has differentiated members to forage, defend, and nurture

performed at unexpectedly low physical levels, since it is to a large extent the paint of the airplane, which is involved in absorbing electromagnetic radar signals at the level of materials engineering. In fact, the electromagnetic absorption occurs at the molecular and atomic level, with quantum mechanics governing the phenomenon.

3.6.3 Trade-Offs

Trade-offs exist in functional allocations. For example, reliability can be enhanced by increasing subsystem redundancy. In particular, the space shuttle has backup mission computers. Reliability can also be increased by distributing functionality—that is, providing that more than one subsystem can perform a crucial function at a critical time. For example, vehicle deceleration can be provided by brakes or by engine braking action. However, redundancies can only be provided with increased cost, and distributed functionality may come at the cost of non-coordination or non-specialization of functional performance. Such trade-offs are the responsibility of the system architect, who must balance different stakeholder interests.

3.7 Modeling Formalisms and Language Development

The use of modeling elements in different architecting area is shown in Table 3.6. Note that the operational/behavioral and the functional/logical architecting areas, or design space dimensions, emphasize the functional analysis and architecting.

Table 3.6 Architecting Areas, Modeling Elements, and Methodologies

Architecting Area	Modeling Elements	Methodologies
Operational/behavioral	Operational processes and support	Decompositions, flows, etc.
Functional	Functions, procedures and facilitators	Decompositions, flows, etc.
Logical	States, conditions, data, and information	Information flows, etc.
Components	Objects and interfaces	Object-oriented design

3.7.1 Modeling Developmental History

A historical outline of system modeling diagrams is show in Table 3.7.

Before object-oriented methods were developed, analytic diagrams tended to either emphasize functions, states, objects, or other defined element types. Object-oriented methods heightened the use of objects, at the expense of functions and states. The unified modeling language (UML) introduced and emphasized the use case as an abstract depiction of system functionality under an operational scenario and as a way to perform functional analysis. Use case emphasis on functionality sometimes means that new or unforeseen functional requirements are discovered, and therefore, the necessity for new subsystems or components is derived. One of the

Table 3.7 Diagrams in Modeling Approaches

Diagrams	Behavioral	Functional	Physical
Traditional Structured Analysis, 1980	Schematic block	Functional flow	Architecture block
Modern Structured analysis, 1988	State (performance specification)	Data flow	Software-product-hierarchy
Object-Oriented Programming, 1990	State	Data flow	Object/class
Unified Modeling Language (UML), 1995	State sequence collaboration	Use case activity	Object/class component deployment
Systems Modeling Language (SysML), 2003	State sequence	Use case activity	Package internal block Block definition

main advantages in using UML is that as you go lower in hierarchical layers, consistency checks are possible at different levels of the functional requirement model.

The advent of the system modeling language (SysML) has balanced the proportional inclusion of different modeling element diagrams. People who are not systems engineers often have a tendency to focus on objects instead of more abstract attributes (Smith and Bahill, 2010), and may be less aware of functions and states as they examine and comment on an engineering design.

3.7.2 Natural Language Parts-of-Speech

For clarity, needs should be written as a declarative statements (e.g., customer needs production capacity), and requirements should usually be imperatives (e.g., *shall* produce) or declarative statements (e.g., production will be 100 t). Functions should be verbs (e.g., produce), subsystems/components/objects should be nouns (e.g., product), and states described with verbs (e.g., producing). Attributes of subsystems/components/objects should be nouns (e.g., productiveness), and attributes of functions should be adverbs (e.g., *efficiently* produce). Table 3.8 shows the correspondences between SE elements and natural language elements.

Erroneous mixing and misuse of natural language parts-of-speech is a plague from which SE is not yet free. SE should be rigorously checked for consistency in the space of associated parts-of-speech. Mismatching of elements produces confusion, which is difficult to remove without correction and reference to fundamental correspondences between SE elements and parts-of-speech.

Table 3.8 SE Elements with Corresponding Natural Language Parts-of-Speech

SE Element	Natural Language Element	Example
Needs	Declarative statement	Customer needs aerial transportation
Requirement	Imperative phrase or declarative statement	Airplane *shall* attain 700 mph Top speed will be 700 mph
Function	Verb	Fly
Subsystem/component	Noun	Wings
State	Verb	Flying, *is grounded*
Attribute of subsystem or component	Noun	Reliability, sturdiness
Attribute of function	Adverb	Efficiently, reliably

3.7.3 Structured Analysis Example

Figure 3.19 shows a hierarchical model of requirement types, using modeling elements of UML and SysML to explicitly depict relations among elements. The figure is an introduction to the generic use of SysML. Figure 3.19 can be explained with the following descriptive phrases:

> Requirement ***has Types***: Functional Requirement, Nonfunctional Requirement, and Composite Requirement (among possible others). Or alternatively,
> Functional Requirement, Nonfunctional Requirement, and Composite Requirement ***Inherit properties from*** Requirement

3.7.4 SysML Diagrams: Requirements, Behavior, and Structure

Logical architecting principles are now best represented by the diagrams of SysML and its parent UML. The diagrams of SysML have undergone extensive vetting and have been determined to be the most efficient set of logical representations for systems generally. Object-oriented methods of modeling have evolved in the UML and its predecessors, until becoming available in the broadly applicable and standardized SysML. SysML diagrams are shown in Figure 3.20.

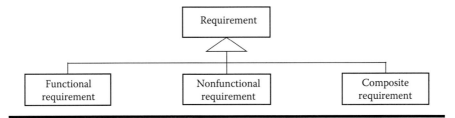

Figure 3.19   Structured analysis of requirements.

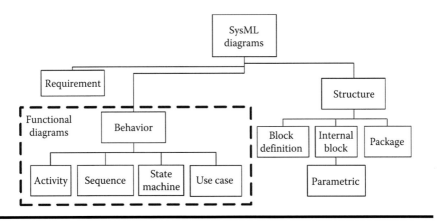

Figure 3.20   SysML diagrams.

The diagrams of SysML are separated into three main groups:

1. **Requirement diagrams** contain listings of imperative requisites that range from customer needs to system specifications. Functional requirements, as an example, can be found in requirements diagrams.
2. **Behavior diagrams** contain functions as main elements, but include other aspects such as time order, states, actors and scenarios, etc. Functions in behavior diagrams should be named and used consistently across all behavior diagrams.
   a. Activity diagrams depict functions as activities. Activity diagrams in SysML combine FFDs with logic and context.
   b. Sequence diagrams depict functions both as commands that signal the next component/subsystem to begin acting, and as the internal actions of the components/subsystems.
   c. State machine diagrams depict functions as actions and activities that maintain a steady condition until a trigger changes the condition.
   d. Use cases depict functions in the actions performed by the actors. Also, use cases may have functional requirements specifically listed in the specific requirements section (Daniels and Bahill, 2004).
3. **Structure diagrams** describe the physical architecture. Structure diagrams do not include functions, unless behaviors are allocated to the structures.

3.7.5 Allocation of Functions and Data to Subsystems or Components

Allocation of functions to components/subsystems is clarified by the development of a coherent set of SysML diagrams, particularly behavior diagrams. The allocation of functions is accompanied by the logical allocation of data to components/subsystems. Logically, functions and the data on which the functions operate should be allocated to the same subsystems or components for reasons of modularity, accessibility, and security, among other important attributes.

3.7.6 Interface Definition: Inputs, Outputs, Ports, and Signals

Interface definition is crucial in SE practice, and SysML tools allow the interfaces to be designed so that integration is subsequently efficient. Inputs, outputs, signals, and ports, among other interface items, should be examined carefully during design, not at the time of integration.

3.7.7 Logical Implementation of Behavior in Subsystems or Components

SysML diagrams, shown as taxonomy, show the variety of representations that are necessary to assure that all essential logic is implemented within systems. Note that the gap between the SysML behavior and structure diagrams is filled by functional allocation and good architecting practice in artfully mating behaviors and structures. As new behaviors are discovered, and as new structures become available with the advance of technology, designs and design patterns must evolve.

3.8 Vasa Case Study

The Vasa was a seventeenth-century Swedish wooden gunship (Ohrelius, 1962) that suffered from a failure in SE practice. Its design flaws will be exposed here in light of our modern understanding of functional analysis and architecting. A distinction between product and process will be drawn. The analysis will first examine the Vasa as a physical product, and then examine the processes associated with the Vasa.

3.8.1 Vasa as a Product

No SE process was followed in the design and construction of the Vasa as a product. Shipbuilding at the time was at the direction of a few knowledgeable shipwrights. In the case of the Vasa, the master shipwright was ordered to change the Vasa's design while it was being built. Although the simple waterfall model is now considered obsolete in SE, even a rudimentary serial design process, such as the waterfall model shown in Figure 3.21 could have prevented the sinking of the Vasa.

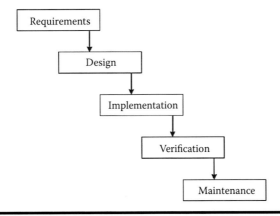

Figure 3.21   Serial waterfall design process that could have saved the Vasa.

The waterfall model indicates that, once implementation is underway, no new requirements or design changes should be undertaken.

3.8.2 Requirements Analysis and the Design Space

No formal requirements were identified or known for the Vasa (NASA, 2008). The design space was entirely inside the mind of the shipwright—a state of affairs that became a central cause of the Vasa's eventual failure. Although master shipwrights were able to built good ships if specifications and decisions were left entirely to their discretion, the king as customer had the power to change requirements during construction, at his pleasure. The problem with the Vasa design and construction process was that the requirements and design were arbitrarily changed by King Gustav as implementation was in progress, and the parts of the ship that were already implemented were conserved and mated with redesigned parts of the ship. Most notably, the one-gundeck ship design was ordered to accommodate two gundecks, making it top heavy and contributing to its sinking.

3.8.3 Conceptual Design and Functional Decomposition

A valid conceptual design is achieved when a functional analysis concludes that the functional decompositions and functional flows are sufficiently realistic, as abstract models. At this point, a valid functional baseline has been modeled. A military ship's top-level functional decomposition could look like Figure 3.22. The *floating* and *sailing* functions have been decomposed one level further than the *fighting*

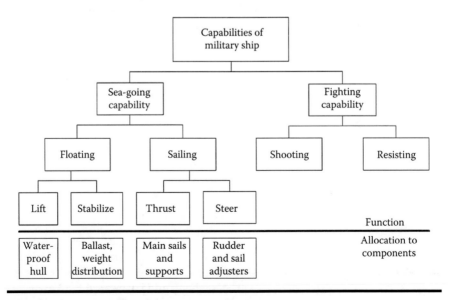

Figure 3.22　Functional decomposition of Vasa as a product.

sub-functions, *shooting* and resisting, since the lifting and stabilizing are highly pertinent to the sinking of the Vasa (NASA). The multiplicity of functions at the top level of this functional decomposition diagram is indicated by the terminology "capabilities," which implies many functions.

The persistent functional nature of a system can be described by a seemingly timeless functional decomposition, with state-like descriptions, that is, float*ing* and sail*ing*. Functions are specified, but there is no indication of when they occur. In this case, the decomposed functions were allocated to physical components in this manner:

Sea-going capability: Physical architecture of the whole ship
Floating: Hull shape and features
Lift: Waterproof hull of the ship
Stabilize: Ballast and ship weight distribution
Sailing: Masts, sails, and rigging
Thrust: Main sails and supporting structures
Steer: Rudder and mechanisms for sail adjustment

A central cause of the Vasa failure was an over-implementation of the shooting function, with a simultaneously deficient implementation of the stabilizing function in the form of ballast and weight distribution. This functional architectural description is complete as an abstract explanation of the physical sinking of the Vasa, but it falls short of describing the essence of the failure of the Vasa, which involved serious flaws in the processes of designing and building the ship.

3.8.4 Logical Architecture

The failure of the Vasa can be explained as the consequence of the ballast and weight distribution throughout the ship, and the shape of the hull. The Vasa was also top heavy because of its elaborate ornaments, but especially because of its extra gun deck. Although modern mathematical ship design formulas make explicit the logic necessary for ship stability and other functions, these formulas were entirely unavailable in seventeenth-century Sweden.

3.8.5 Physical Architecture

Since the Vasa designers did not assure the performance of necessary functions through the logical design of the Vasa, the physical architecture was severely flawed. The physical configuration of the Vasa could not perform its functions, principally, to stabilize, and the Vasa subsequently sank.

3.8.6 Design and Construction Processes for the Vasa

A root cause of the tragic sinking of the Vasa as a product was the improper execution of the processes involved in designing and building the Vasa.

3.8.7 Functional Decomposition of Processes

Figure 3.23 shows the functional decomposition of processes associated with Vasa's design and construction.

Traditionally, a master shipwright had the authority to both design and lead the construction of a ship. In the case of the Vasa, however, the functional activity of designing the ship was improperly performed by the king, beside the shipwright. This violated the most fundamental principal of functional allocation, which is that each function should be allocated to only one subsystem, or person. This principle, in the domain of jobs and people, is the principal behind the usual sole ownership of a job or responsibility.

"Designing and constructing the Vasa" is a process that could only have been performed well if the functional flows were well-defined and stable. The functional flow should have included design, engineering, construction, oversight, and review activities. In addition, the functional flow should have been respected by the parties involved.

Note that the scope of the design and construction process is limited for understandability. Thus, only a top-level FFD of this process appears in Figure 3.24.

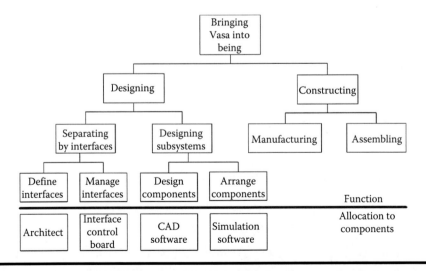

Figure 3.23 Functional decomposition for "Bringing the Vasa into being."

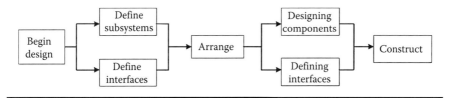

Figure 3.24 Functional flow of "designing and constructing the Vasa."

3.8.8 Logical Architecture of Vasa Design Process

Logic can be added to the functional flows in order to describe decision and oversight features of the design and construction process. This is shown in Figure 3.25.

The physical architecture of the design and construction process is dynamic, transitory, and non-permanent. Specifically, the physical nature of the design and construction process involves workers properly moving in space and time in order to fulfill their job functions.

3.8.9 Operations: Sailing the Vasa

Decomposing the functions involved in sailing the Vasa and actually using it in warfare is the first step in defining the operational functional flows of the Vasa. For example, we can decompose the function of tacking, or sailing upwind in a zigzag pattern, into the functions of *steer* ruder and *adjust* sails. At this point, the logical details of the relationship of these two sub-functions will begin to emerge. Then, the logical architecture of sailing multi-masted ships could be researched and called upon to inform an engineer about the logical architecture of sailing the particular ship, that is, the Vasa.

The sub-function, adjusting sails, is itself a complicated process, involving the coordinated adjustment of many sails, which could be described with logical details. For example, sailors use visual cues as data, and pass messages among themselves,

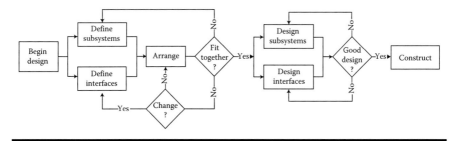

Figure 3.25 Logical architecture of designing and building the Vasa.

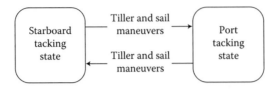

Figure 3.26   State diagram: sailing upwind by tacking.

in order to choreograph the lowering, raising, changing, and adjustment of sails as part of the fine-tuned process of making best use of the winds available. There are many other logical details associated with the process of sailing a ship. Some of the logical details are shown in the state diagram in Figure 3.26.

3.9 MMS Case Study

3.9.1 Service Decomposition

In this section, we analyze modern communications services first from the generic view of an end user sending information to any other end user that subscribes to a similar service. We then qualify the communication services by the type of information that can be exchanged among users: first voice services only and then multimedia services (MMS) (voice, fax, data, video, and/or multimedia). Figure 3.27 shows the generic situation with top-level management services and operations, as well as the functions, access and transport, allocated to separate networks.

3.9.2 Functional Decomposition

We start with the following mission statement: "The system shall provide information services to anyone, anywhere, anytime." It is clear that to send information, we

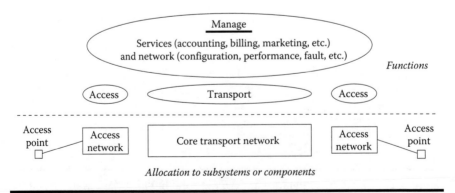

Figure 3.27   Generic view of multimedia services.

need some kind of access and transport function to transmit the information from originating point A to a terminating point Z. Also, to provide customized service, we need a manage function to know who the customer is, and to which services he or she has subscribed. Note that we concentrated on "what" the system must do (functions) and not "how" the system will do it. A functional decomposition diagram, with functional allocation to subsystems, appears in Figure 3.28.

Following the methodology discussed in the above sections, we have defined the following sub-functions:

F0: Provide multimedia services
F1: Access
F2: Transport
F3: Manage

From an information services perspective, these functions are independent of the type of service.

3.9.3 Classical TMN Architecture

A detailed functional decomposition has been worked out over the years and there are standards like the International Telecommunication Union (ITU), *telecommunications management network* (TMN), and the *enhanced telecomm map* (e-TOM) for business processes of service providers. As an example, the TMN standard defines the functions in the following layers:

1. Business management layer
2. Service management layer
3. Network and systems management layer
4. Element management layer
5. Network element layer

These layers are shown in Figure 3.29.

3.9.4 Functional Flow and Logical Architecture

Figure 3.30 depicts the top-level functions of a communications system. The request service function is the initial end-user request (subscriber to the service) to set up an information exchange session (a voice call, facsimile transmittal, text message, e-mail, video, etc.). In traditional twentieth-century networks, the access function looks at the request to determine destination point, to determine available resources to route to the core network. The transport function looks at the addressing and routing of the call to its final destination, core network resource allocation (bandwidth), address translation, egress node resource

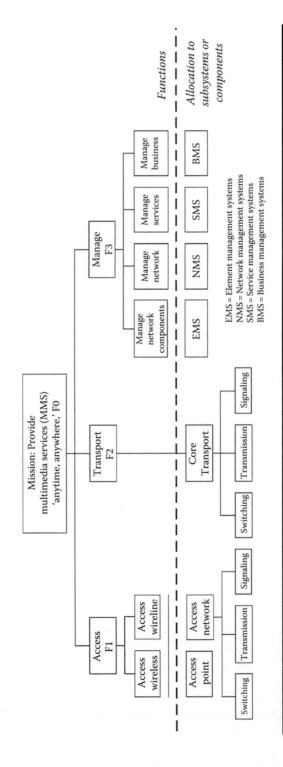

Figure 3.28 Mission, services/operations, and subsystems.

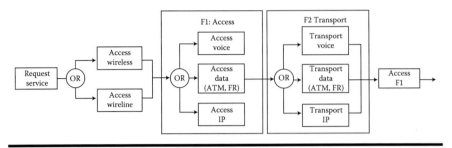

Figure 3.31   Twentieth-century functional flow.

This functional decomposition is basically the same as before, but the operations (behaviors) and the allocation to physical architecture components are different; for example, the access function was allocated to voice through voice switches, data through frame relay or ATM switches, and Internet Protocol (IP) through IP routers.

In today's environment, the architectures that are being defined, developed, and deployed are open architectures using the IP as the common transport platform. This is the objective for the next generation networks (NGNs) that implies functions detached from the network elements for admission control, multimedia control, class of service, core network resource assignment, end-user authentication, etc. So, we need to add a function F4 control, which will have F4.1 control network, F4.2 control service, and F4.3 control applications. Figure 3.32 shows the expanded functional decomposition in NGN, and Figure 3.33 shows the expanded functional flow. This will become clearer when we discuss the TLD for call setup in a classical voice and a MMS network in the following section. This separation of functionalities can be better understood by analyzing the NGN standard for the functional architecture of MMS shown in Figure 3.34.

3.9.5 Timeline Diagrams

3.9.5.1 Voice Call Setup

Figure 3.35 is a high-level representation of the functional sequence in the establishment of a traditional voice call setup.

A service request happens when the end user lifts the hand set; access node (voice switch) acknowledges (ACK) the request by providing a dial tone to the end user to start dialing the destination number; then, the access node looks at the digits dialed and determines destination and makes digit translation to send the initial address message (IAM) to the network node. The network node in turn does digit analysis to determine the routing nodes and the destination node at which the called party resides. If the egress node determines that the egress phone is idle, then it will alert the originating party and also will send a pulse for ringing

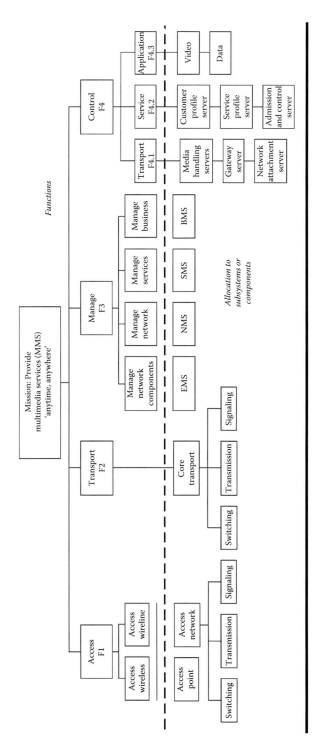

Figure 3.32 Expanded functional decomposition.

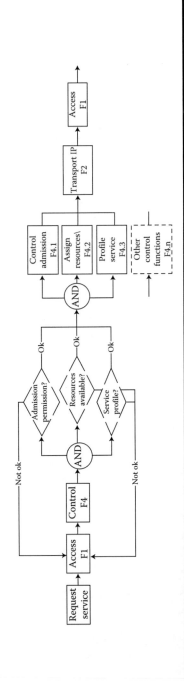

Figure 3.33 Expanded functional flow diagram.

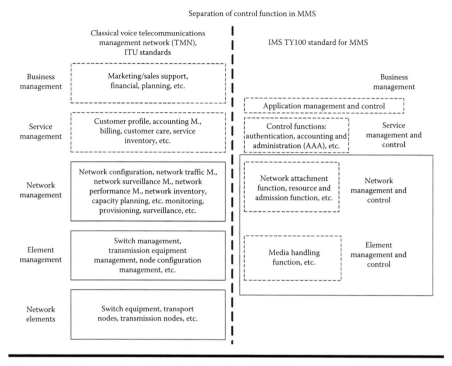

Figure 3.34   Separation of control functions in MMS.

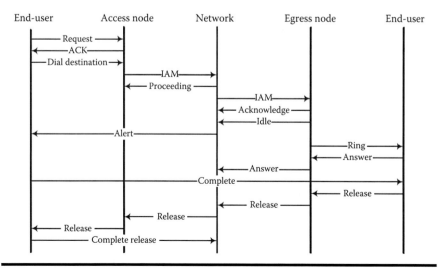

Figure 3.35   Functional timeline for voice call setup.

at the terminating (egress) end. When the incoming call is answered, a sequence of answer messages is sent through the different nodes to close the circuits to establish the voice path for the call.

3.9.5.2 Multimedia Session (MMS) Setup

In a multimedia network, we talk about a session establishment where, as mentioned before, the session may be voice, facsimile, data, video, etc. Figure 3.36 show a high-level timeline diagram for session establishment.

The initial request is through an INVITE message; the access node sends the request to an access server to authenticate and check for authorization privileges of the end user; if authorization and authentication is successful, then the access node will send an ACK message and also will check with a network server for available network resources and also check with the application server to see to what applications the subscriber has subscribed. Once the service profile and the network resources are identified, an INVITE message will be sent to the terminating party to join the session; if the invitee accepts, then a series of messages are exchanged to establish the session (CONNECT) and the session is established.

This example shows that some of the functions that had been allocated to the network before are now allocated to servers that will instruct the network on the resources (bandwidth, quality of service, etc.) that need to be reserved for the session to be established. The core network in this case has a pure TRANSPORT function, which is one of the main drivers of the NGNs.

3.9.6 Physical Architecture

As seen in the last section for the *information services* example, there were separate networks defined for different type of services; this is depicted in Figure 3.37. The services were vertically built on top of application-specific TRANSPORT functions (e.g., voice network, frame relay, ATM, Internet, etc.). Taking a closer look at the physical architecture, we assign physical nodes with its corresponding functions to provide the service. Then, we have *voice switches* (voice SW) and its corresponding traffic network (e.g., dedicated bandwidth in the access and core transport to voice traffic), *frame relay switches* (FR SW) and its corresponding FR traffic network (e.g., dedicated bandwidth in the access and core transport for FR; IP SW, etc. This is shown in Figure 3.37.

In the traditional network, there were also functions that were allocated in a hierarchy for interconnection among different regions; the local functionality was assigned to local switches, and then there are gateways to interconnect different local serving areas, which in turn will also interconnect to regional nodes to allow traffic among different regions (e.g., at the national level) and nodes, which will have the needed functionality for international interconnection. The Figure 3.38 shows the nature of the hierarchical network.

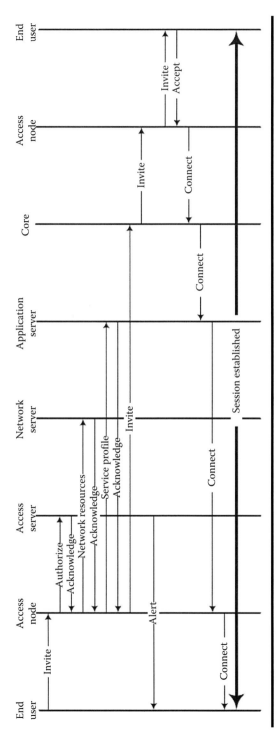

Figure 3.36 Timeline for a MMS setup.

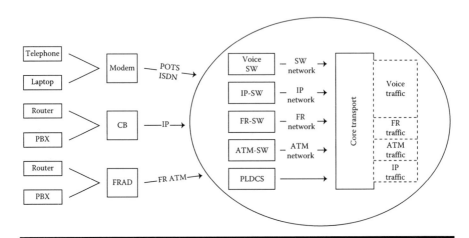

Figure 3.37   Twentieth-century architecture.

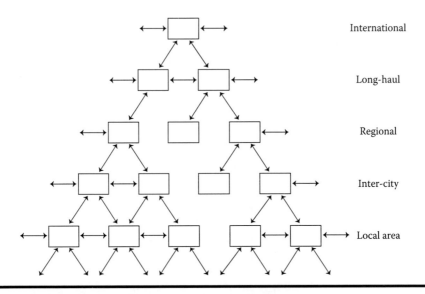

Figure 3.38   Physical architecture hierarchical network.

In NGN, the layered physical architecture allows any service delivery via any of the wireline or wireless technologies available today: digital subscriber loops (DSL), hybrid fiber coax (HFC), fiber (fiber to the curb "FTTC," etc.), wireless point-to-multipoint (LMDS, MMDS, etc.), WiMax, Wi-Fi, and third generation (3G), fourth generation (4G), or NGN. The transport layer is pure IP network with different types of technologies available (DWDM, gigabit ethernet switches (GbE),

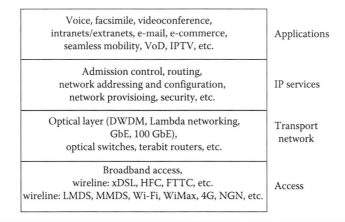

Voice, facsimile, videoconference, intranets/extranets, e-mail, e-commerce, seamless mobility, VoD, IPTV, etc.	Applications
Admission control, routing, network addressing and configuration, network provisioing, security, etc.	IP services
Optical layer (DWDM, Lambda networking, GbE, 100 GbE), optical switches, terabit routers, etc.	Transport network
Broadband access, wireline: xDSL, HFC, FTTC, etc. wireline: LMDS, MMDS, Wi-Fi, WiMax, 4G, NGN, etc.	Access

Figure 3.39 MMS architecture framework.

optical switches, terabit and/or gigabit routers, etc.), which transport buckets of data to its final destination.

There are still very interesting discussions in the technical forums about the best way to evolve this transport layer (IP over the optical layer, pure IP, IP over SDH+, etc). Figure 3.39 shows the services layer and the applications layer consists of servers that contain the logic and functions needed to establish the MMS session.

3.10 Conclusion

Functional analysis is central to design, as has been recognized as such by modeling methods from classical structured analysis all the way to the SysML and its behavior diagrams. An organized, methodical approach to functional analysis, which gives attention to the SE process, important modeling elements and methods, functional and logical analysis principles and processes, and physical architecting principles, is the core of design. Adequate knowledge and tools enable the correct decomposition and allocation of functions—the simplest way of describing the design process. A study- and practice-based approach to gaining design mastery is ultimately based on case studies and the wisdom they impart to the SE analyst and architect.

Authors

Ricardo L. Pineda holds PhD and MSc degrees from Lehigh University and a BSc degree from Universidad Nacional de Colombia. He has over 20 years of experience in systems engineering (SE) in different industries ranging from research and development at Bell Labs to chief technology officer at AT&T in Mexico. He was a distinguished member of the technical staff at Bell Labs where as a systems engineer he worked on requirements and the architecture of new technologies used in the AT&T network. He was the chief technology officer for the definition, development, implementation, and deployment of the AT&T network and services in Mexico. He was also practice director of Siemens Business Services for Latin America where he was the main consultant in systems implementations in Venezuela, Colombia, Ecuador, and Brazil. Dr. Pineda has extensive experience in academia; he was a professor at ITESM in Monterrey, Mexico, and at the "Universidad de Los Andes" in Colombia. His current research projects include PI for "Prognosis & Resilience Design for Complex SoS" with Raytheon-IDS, PI "SOS Global Attributes to Design Space Mapping" and "Technology Refreshment Assessment Model" for LMC-Aero, PI "Level of Repair Analysis Models" for Hamilton-Sundstrand, and PI for the "TMAC El Paso del Norte-Region" for the MEP program sponsored by NIST. Dr. Pineda was nominated to receive the "US-Hispanic Engineer of the Year Award," received the "AT&T Architecture Award," and was honored with the "Baldwin Fellowship" and "Gotshall Fellowship" awards.

Currently, he is at the University of Texas at El Paso where he is the director of the Systems Engineering Program, the director of the Research Institute for Manufacturing and Engineering Systems, and the chair of the Industrial, Manufacturing and Systems Engineering Department. He is also a member of International Council on Systems Engineering, IEEE, and ISACA.

Eric D. Smith is currently an assistant professor at the University of Texas at El Paso (UTEP), working within the Industrial, Manufacturing and Systems Engineering Department and its Systems Engineering Program, as well as within the Research Institute for Manufacturing and Engineering Systems (RIMES). He earned a BS in physics in 1994, an MS in systems engineering in 2003, and a PhD in systems and industrial engineering in 2006 from the University of Arizona in Tucson, AZ. His dissertation research lay at the interface of systems engineering, cognitive science, and multi-criteria decision making. He taught for 2 years in The Boeing Company's Systems Engineering Graduate Program at the Missouri University of Science and Technology. He has given invited talks at The Boeing Company, on the topic of risk management, and for a multiple university research initiative (MURI) composed of the Universities of Arizona, Arizona State University, Ohio State University, the University of Florida, and the University of Michigan, on the topic of ameliorating mental mistakes in uncertain and complex environments. Currently, he works with Lockheed Martin Corporation's summer project practicum for systems engineering students. His current research interests include complex systems engineering, risk management, and cognitive biases. He is a member of INCOSE, ITEA, IIE, ASEM, and ASEE.

References

Bahill, A.T., Szidarovszky, F., Botta, R., and Smith E.D. 2008. Valid models require defined levels. *International Journal of General Systems* 37(5): 553–571.

Beer, S. 1985. *Diagnosing the System: For Organizations*. New York, NY: John Wiley & Sons.

Bertalanffy, L.V. 1933. *Modern Theories of Development: An Introduction to Theoretical Biology*. London, U.K.: Oxford University Press.

Bertalanffy, L.V. 1968. *General Systems Theory: Foundations, Development, Applications*. New York, NY: George Braziller.

Bertalanffy, L.V. 1981. *A Systems View of Man*. Boulder, CO: Westview Press.

Daniels, J. and Bahill, A.T. 2004. The hybrid process that combines traditional requirements and use cases. *Systems Engineering* 7(4): 303–319.

INCOSE. 2006. *Systems Engineering Handbook: A Guide for System Life Cycle Processes and Activities*. INCOSE-TP-2003–002–03, version 3. Seattle, WA: International Council on Systems Engineering.

Mil-Std-499A. 1974. *Military Standard: Engineering Management*. United States Department of Defense.

NASA. 2008. Lauching the Vasa case study, National Aeronautics and Space Administration.

Ohrelius, B. 1962. *Vasa: The King's Ship*. London, U.K.: Cassel & Co.

Smith, E.D. and Bahill, A.T. 2010. Attribute substitution in systems engineering. *Systems Engineering* 13(2).

Chapter 4

System Verification, Validation, and Testing

Ricardo L. Pineda and Nil Kilicay-Ergin

Contents

Verification, validation, and testing (VV&T) has been subject to varying interpretations ranging from software code checking to a more complex set of processes that take into consideration the full life-cycle development phases of systems. VV&T is a set of processes, tools, and analytical techniques used for the detection and correction of system flaws ultimately leading to reduced risk of system failures and higher customer satisfaction. Conventional method of performing VV&T late in the system development phase results in late detection of errors and substantially increases project costs, delays delivery schedule, and introduces significant risks. As contemporary engineered systems are becoming more integrated and capability driven, VV&T becomes a complex and vital process that requires a systematic approach.

The main focus of this chapter is on a series of technical and management activities necessary for successful VV&T of complex systems. This chapter discusses VV&T methodologies and tools for complex systems, which consist of hardware components, software components, and processes. Architectural attributes that affect the VV&T process, verification methods, verification planning and implementation, verification phases throughout the system development cycle, system testing planning and deployment, as well as validation are discussed in detail. A classical virtual network service (VNS) of the late twentieth century is introduced as an example of VV&T processes. This VNS is then compared to current plans for the multimedia networks comprising the next generation networks (NGNs). These examples show the evolution of VV&T methodologies and processes in the fast-changing environment. Finally, a case study on the current NASA Constellation program is presented, which illustrates VV&T planning in the aerospace industry.

4.1 Introduction

System design and development is a complicated and diverse process, which depends on the combined cognitive strategies of the people involved in it. The more complex developer interactions get, the more problems. For example, problems are inevitable when 200–3000 cross-disciplinary team members in multiple sites have to interact at many levels of communications. There are well-established project management strategies and methodologies to deal with these problems. However, as the customer expectations for system performance continue to increase, the customer requirements for system development tend to become vague and fuzzy. This leads to challenges in terms of satisfying customer expectations. Verification, validation, and testing (VV&T) is a set of processes and activities to reduce uncertainty in systems by identifying useful information on stakeholder expectations. The objective of the VV&T is to identify problems in requirements, design, production, and maintenance phases of system development. Therefore, VV&T is a vital process to enhance the technical success of contemporary complex system development.

Life cycle models describe systems from the beginning of product development to the end of disposal. Different government organizations and industries define various life cycle models. The V-model (*INCOSE Systems Engineering Handbook*, v3, 2006) focuses on the systems development stage and summarizes the product development phases along the verification and validation phases. The V-model is effective in placing the VV&T activities in the context of the system development life cycle (Figure 4.1). At each stage of the decomposition process, relevant requirements constraining the design solution space are identified and for each requirement, it is necessary to define the verification method that will prove that the design solution will satisfy the requirement.

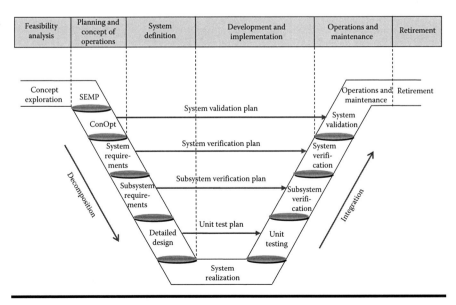

Figure 4.1 V-Model—VV&T in context with system development life cycle.

As shown in Figure 4.1, VV&T is entangled with the integration of components and subsystems. Therefore, VV&T complexity and quality are directly related to system integration (SI) complexity. Most of the factors affecting SI complexity also impact the quality of system verification and validation (V&V). A survey across government organizations and industry sectors has been conducted to identify the factors that affect system architecture design and eventually affect SI and system V&V quality. Based on the survey, major architectural factors that affect V&V quality are identified as follows (Jain et al., 2008):

1. Identification of clear system requirements: Requirements engineering plays an important role in VV&T quality. Reducing the ambiguity and volatility of requirements as well as focusing on traceability across all phases of the system development improves the VV&T quality.
2. Requirements prioritization: Prioritization of requirements supports iterative development process and if this activity is performed early in the life cycle with stakeholder participation, VV&T can be planned and executed more effectively.
3. Modularity: Physical and functional modularity in architecture facilitates replacement of legacy systems, which reduces the complexity. This moderately improves VV&T quality.
4. Operational commonality: Commonality is the extent to which the system is made up of common hardware and software components. Commonality in system architecture results in operational requirements that can be easily verified and validated.
5. Commonality in subsystems: Reducing unique interfaces, platforms, and protocols results in familiarity of subsystems, which reduces the activities required for integration, verification, and validation.
6. Familiarity of technology: This factor reduces the configuration complexities of SI process and can improve V&V moderately.

We can understand the VV&T challenges and constraints for contemporary systems by looking at how these architectural factors are shaped in contemporary systems. Contemporary systems such as system of systems (SoS) and network-centric systems (NCS) are formed through evolutionary development, and thus requirements are ambiguous and fuzzy. Besides, the systems that constitute SoS operate independently and have different local goals resulting in reduced operational commonality. There is a significant increase in integrating currently available systems into SoS, which brings along the challenges of integrating legacy systems. The success of creating modern systems depends on overcoming these challenges through systematic VV&T processes. The goal of this chapter is to outline the necessary VV&T for successful system realization. The rest of the chapter elaborates on the VV&T methods, processes, and types across the system development. VV&T for two applications are discussed to provide more insights. The VV&T challenges for contemporary systems such as NCS and SoS are discussed in the context of these application domains.

4.2 Verification

Verification is the process of confirming that the system and its elements meet its requirements. It ensures that the system performs its intended function and meets performance requirements allocated to it. *NASA Systems Engineering Handbook* (NASA/SP-2007-6105, Rev 1) defines two major verification types: requirements verification and design verification. Requirements verification is conducted as an essential part of SI where a preliminary strategy for how a requirement will be verified is agreed at the initial systems requirements review. This preliminary verification strategy forms the basis for the establishment of verification requirements and plans that are created after the completion of the critical design review. Design verification focuses on evaluating the design including hardware, software, processes, and people to ensure that the requirements have been met. The design verification is accomplished by verification methods: analysis, test, inspection, and demonstration.

During design verification, the design is assessed recursively throughout the project life cycle to determine if the system is built right. This can be performed on a wide variety of product forms such as (Grady, 2007)

1. Algorithmic models, simulation models, and virtual reality simulators
2. Mockups
3. Conceptual models
4. Prototypes with partial functionality
5. Engineering units with full functionality but partial form or fit
6. Test units with full functionality, form/fit but partially integrated
7. Qualification units that can be subjected to extreme environmental conditions
8. End product that is ready for full functionality or service

These product forms can also be in different states such as produced (built, fabricated, manufactured, or coded), reused (modified internal nondevelopmental products or OTS product), and assembled and integrated (a composite of lower-level products) (NASA/SP-2007-6105, Rev 1).

4.2.1 Verification Methods

Test: Testing is a dynamic verification method where established scientific principles and procedures are applied to determine the performance or properties of the product under test. In some cases, special test equipment may be necessary to quantitatively measure the performance of the product under test. Testing is the most data-intensive verification technique and can be conducted on final end products as well as prototypes. There are many types of testing: environmental stress testing, structural testing, node testing, network testing, operational testing, manufacturing testing, safety testing, security testing, black-box testing, and white-box testing.

Analysis: Analysis is a verification method where established analytical techniques, mathematical models, simulations, algorithms, graphs, or other scientific principles are applied to determine whether the specified requirements are met. It is used intensively during conceptual and detailed design phases when the end product, the prototype, or the engineering model is not readily available.

Demonstration: This is a method that checks the actual operation, adjustment, or reconfiguration of products to determine whether required functions are met. The product under demonstration is operated under specific scenarios and performance is monitored on a qualitative basis. The goal is to observe whether the product under demonstration satisfies required operating characteristics. When compared to testing method, demonstration confirms performance without intensive data gathering.

Examination/Inspection: This is a static method where observations are made in a static situation without utilizing special procedures. This method encompasses the use of the five senses, simple physical manipulation, mechanical/electrical gauging, and measurements. It can be in the form of a review where software code is inspected without execution or it can include the review of documentation and comparison of product/item's physical features with predetermined standards.

4.2.2 Verification Types/Classes

One of the main objectives of verification is to reduce uncertainty and risk by identifying problems in all phases of the system design development. DoD and NASA defined verification classes (types) based on the life cycle phases and level of the end product within the system. From conceptual design to detailed design phases, end products that have not been used in a particular application before are verified for that application. Then, at the production/construction phase, every end product produced, which is intended for delivery to the customer, is verified. Finally, prior to system utilization phase, the end product is operated under various scenarios to verify the product's operational capabilities (Levardy et al., 2004). These three verification types based on life cycle phases are defined as following:

Qualification Verification: This verification type focuses on confirming end products that have not been used before. This is especially important for end products that require safety in terms of danger to property or loss of life. Therefore, measurement tolerances are tighter compared to other verification classes. Every verification activity is driven by specifications. In the case of qualification activity, performance specification prepared at the end of conceptual design phase is used to drive qualification verification. Qualification verification plan is created based on this specification, which outlines what is required from the system without specifying how to do it by limiting system developer to specific materials, processes, or parts.

Acceptance Verification: This verification type is conducted on each end product that is intended for delivery to the customer and provides proof of acceptability to the customer. Since it is intended for delivery to the customer, the acceptance measurement tolerances are wider compared to qualification verification. Acceptance can be conducted in two consecutive steps: (1) article verification and (2) recursive verification of every article manufactured on the actual production line. Acceptance verification activities are driven by the detailed design specification (product specification) that describes actual features and functional characteristics of the product. The acceptance verification plan is created based on this specification. The products that pass acceptance verification are intended for delivery to the customer but the products must pass a final verification step prior to customer use.

System Test and Evaluation: It is not possible to analyze every combination of system operation analytically. It is necessary to conduct actual trials to observe the actual behavior and performance of the final system. System test and evaluation is a type of verification where the final end products are operated at the system level to confirm that the system satisfies system requirements. Even if all the subsystems and components have passed the qualification and acceptance verifications, the whole has to be tested as well. This is especially vital in complex systems such as space systems. This verification method is discussed in Section 4.4 in more detail.

Another way to look at verification classes is that verification is conducted on a wide variety of end products that are at different levels of the system structure. Based on that perspective, unit testing is conducted on end products that are the lowest-level components of the system. Once all units are verified, the verification transitions to the next level up, which is called integration or configuration testing. These verification types based on levels of system structure are defined as following (Kaner et al., 1999):

Unit Testing: This activity focuses on verifying the individual units that make up the system. The local structure and functionality of the unit is verified. Also at this level, relevant interfaces are verified to ensure that proper information flows in and out of the unit and the unit is compatible with other units of the system.

Integration/Configuration Testing: This activity focuses on verifying a complete system or subsystem, which is composed of several integrated units or subsystems. It is an incremental approach where the system is configured and tested in small segments.

Regression Testing: Each time a new unit is added as part of integration, the system changes. These changes may cause problems with functions or system capabilities that previously worked flawlessly. Regression testing is the re-execution of a subset of test cases that have already been conducted to ensure that changes do not introduce unintended behavior or additional faults.

Unit testing, integration testing, and regression testing are intensively utilized for verifying software systems or embedded systems. Regardless of the verification class, there are systematic processes for conducting successful VV&T. The following section discusses verification processes.

4.2.3 Verification Process

A well-organized verification process plays an important role in system quality control. There is a direct relationship between VV&T and system quality. System quality cost is composed of three main components: prevention cost, appraisal cost, and failure cost (Hoppe et al., 2007). Prevention cost includes any proactive activity cost (i.e., quality assurance and quality training) incurred to reduce the risk of defects. Appraisal cost includes any cost incurred (VV&T cost) to evaluate requirements compliance. Failure cost includes any cost incurred to rework, retest, and redesign. In most cases, the system cost increases drastically due to failure costs identified late in the life cycle of the systems. A systematic verification process and frontloading of verification activities early in the development will reduce the uncertainty and decrease failure and rework, thus reducing system cost. Verification process consists of two main steps: At the verification planning stage, the requirements and constraints that will guide the planning are outlined and the activities, tools, and methods necessary for verification implementation are planned at the beginning of each life cycle phase. During execution stage, the planned verification activities are executed and the results are analyzed to evaluate the system.

4.2.3.1 Verification Planning

A well-planned verification plan saves on budget by reducing or eliminating redundant tests. It is also a control mechanism for program schedule in terms of resources and time. Therefore, verification planning focuses on appropriately organizing activities, methods, and tools to reduce program risk while improving on cost, quality, and development time (Hoppe et al., 2007). Grady (2007) provides a systematic approach to verification planning. The approach can be successfully applied to different types of programs including highly complex programs. This section will adapt and outline the same systematic approach. Interested readers can find detailed information in Grady (2007).

Verification planning starts with defining verification requirements. The right timing for writing verification requirements is at the same time the end product requirements are defined. This encourages writing better requirements as well since verification requirements can only be derived from clear and quantifiable requirements. Once verification requirements are documented, the next step is to create the verification traceability matrix (VTM), also called verification matrix. This matrix is a preliminary plan for how a requirement will be verified. The matrix coordinates the relevant requirement used for the development phase of interest with the verification method that will be applied to confirm the requirement and relates these to the corresponding verification requirement. Table 4.1 illustrates a VTM sample. The sample matrix is created for qualification verification planning and utilizes performance requirements that have been developed at the end of conceptual design phase to drive the verification requirements. This matrix is a useful tool to map related requirements to verification requirements, which forms the basis for designing verification tasks. The ID numbers are important to create traceability among the complete set of documentation. In most cases, this ID number is assigned automatically by the database used for requirements management.

Once all of the verification matrices for all levels of the end product are created, a verification compliance matrix (VCM) is assembled from the union of all verification matrices. Table 4.2 illustrates a sample VCM for qualification verification. The matrix combines unit-level, subsystem-level, and system-level verification requirements. There is an additional column where these verification requirements are mapped to verification tasks. It is important to notice that one verification task can verify several verification requirements at the same time. This is how planning reduces verification cost by reducing or eliminating redundant tests.

The compliance matrix is useful for designing the verification process. The matrix is supplemented with a verification task matrix that defines verification tasks by identifying the principal engineer responsible for the task, the planned start and end dates, the allocated budget for the task, and the related documentation that describes the task procedure in detail. Table 4.3 illustrates a sample verification task matrix. A key point is that all three matrices—verification matrix, the compliance matrix, and the task matrix—are linked together. This enhances traceability and helps manage the evolving processes.

Once the verification task matrix is formed, the task procedures are documented in detail to complete the verification planning and move to verification deployment. The responsible engineer or department produces the planning associated with the task and prepares the procedures. Two or more tasks may require the same resources at the same time or one task may require inputs from another task. The planning should identify these conflicts and resolve them appropriately. Figure 4.2 summarizes the verification planning process.

4.2.3.2 Verification Deployment and Management

Once all verification tasks are identified and documented in the verification planning, the verification deployment phase can start. Every verification task involves preparation for the task, followed by task implementation and finally collecting and analyzing the results. Preparation for the task involves activities to bring together necessary personnel, facilities, resources, raw material, software, and

Table 4.1 Verification Matrix Sample

Performance Requirements			Verification Methods				Verification Requirements	
Requirement Paragraph	Requirement ID	Title	Test	Analysis	Examination	Demonstration	Verification Requirement Paragraph	Verification Requirement ID
3.3.1	123ABC	Motor reliability	X				4.4.5	11ZX79
3.3.2	1AB2C3	Motor reliability				X	4.4.6	5Y6TM

Table 4.2 Verification Compliance Matrix Sample

Performance Requirements			Verification Methods					Verification Requirements		Verification Tasks
Requirement Paragraph	Requirement ID	Title	T	A	E	D	Level	Verification Requirement Paragraph	Verification Requirement ID	Verification Task Number
3.3.1	123ABC	Motor reliability	X				Unit	4.4.5	11ZX79	12
3.3.2	1AB2C3	Motor reliability				X	Subsystem	4.4.6	5Y6TM	12

If the results of the verification task are unsatisfactory, it will be necessary to identify the source of the failure in order to identify appropriate corrective action. If the failure is due to incorrect execution of the task, the task can be repeated. If the failure is due to a design problem that has to be corrected, then the end product will need to be re-verified. In other cases, the failure could be due to an unrealistically demanding requirement, and verification procedures prove that it cannot be fully satisfied. In such cases, it may be necessary to change the requirement in accordance with the change control process previously agreed upon with the customer.

It is also important to manage the planning since plans are predicted activities about the future and involve risks of failure when they are actually implemented. An accident that damages the test equipment may occur, materials necessary to start the task may not arrive in time, or unpredicted weather may delay the planned schedule. Therefore, it is vital to utilize project management tools such as critical path method and Gantt charts to track the progress of planned activities and prepare contingency plans for possible risks that may delay the activities.

It is clear that the verification process is a very data-intensive activity including verification requirements, plans and procedures, and reports of results. Therefore, it is important to utilize effective database tools to support the verification activities and capture the complete set of documentation.

4.3 Validation

Validation is the process of determining whether the system functions in the manner its stakeholders/customers expect when used in its intended environment (Hoppe et al., 2007; Levardy et al., 2004). A system can be verified and meet all its requirements, but this does not guarantee that the system is validated. Therefore, similar to the verification, it is necessary to conduct a series of activities to confirm that end products at every level of the system development process satisfy customer expectations outlined in concept of operations. Customer participation is the key point in validation and it is conducted throughout the life cycle of the system: requirement validation, design validation, and product validation. Requirement validation focuses on requirements and ensures that requirements are understood, concise, complete, and correct. It is conducted before the design. Design validation focuses on the system design and ensures that the design meets the operational needs of the customer. It is conducted before the actual system is built. Product validation focuses on the realized end products and ensures that end products meet the operational expectations of the customer. It is conducted after the end product is verified. This section discusses the validation methods and processes.

4.3.1 Validation Methods

Test, analysis, examination, and demonstration are the types of methods utilized to validate systems. The method names are the same as verification methods, but the objective of conducting these methods is different. Validation is conducted in the intended operational environment whereas verification is conducted under the controlled conditions of the developer.

Test: This method applies well-established scientific procedures to obtain detailed data about the performance of the realized end product.

Analysis: This method applies mathematical and analytical quantitative techniques to obtain proof about the suitability of designs for customer expectations.

Inspection: This method is used to validate the physical features of the design.

Demonstration: This method is used on realized end products under realistic conditions to provide basic confirmation on performance of the design.

Regardless of the validation method used, validation requires a systematic process. The following section discusses validation processes.

4.3.2 Validation Process

The validation process starts with validation planning and proceeds to validation deployment and management. The aim of this process is to ensure that the right system is designed and the system can be used by the intended users. The process also confirms whether the system satisfies the measures of effectiveness defined at the beginning of the program. Any problems identified during the validation process must be resolved before delivery of the system to the end user. Even though the V&V have different objectives, both processes are similar and can overlap.

4.3.2.1 Validation Planning

Validation planning starts with identifying the set of requirements to be validated. Then, the validation methods that will be used for validating the requirements should be defined based on the life cycle of the end product and the level of the end product within the system structure. A similar systematic approach used for verification can be utilized at this phase where validation traceability matrices are created and then combined into a validation compliance matrix from which validation tasks/scenarios can be formulated. As the validation procedures are planned, any constraints, success criteria, resources such as facilities, and operators should also be identified and documented in detail. The validation plan documentation should include validation methods that will ensure that the end product or system satisfies stakeholder expectations as well as validation procedures and a description of the validation environment such as facilities, equipment, simulations, personnel, and operational conditions.

4.3.2.2 Validation Deployment and Management

Validation deployment and management start after the validation strategy is reviewed and approved for implementation. Every validation procedure involves preparation for the validation. The end product to be validated should be available and have already passed verification. All the resources needed to conduct the validation procedure should be available as well. This includes equipment, personnel, as well as facilities in which validation will be conducted. Once validation readiness is approved, the actual validation procedure can be conducted. Data are collected during this stage for required validation measures. At the completion of validation, collected data is analyzed to come to a conclusion regarding passed/failed validation. At the end of validation deployment, end product deficiencies are identified and corrective action reports are generated. Configuration management plays an important role at this phase to update new versions of the reports and disseminate the information to all relevant program participants. If redesign is necessary for the end product, the product will have to be re-verified and revalidated. If the redesign involves software, it is also important to conduct regression tests to ensure that the change did not affect the previously accepted functions.

4.4 System Testing

It is not possible to analyze and exercise every combination of system operation analytically. Besides, there are uncertainties in system behavior such as human operator response to unusual conditions or the combination of several environmental stresses at the same time. Therefore, it is necessary to conduct actual operation of the system in its actual environmental setting. System testing is a subset of V&V, which focuses on evaluating system operation in its actual environment.* System testing cannot be conducted on requirements or the design of a system. It is necessary to have the complete integrated system available. The subsystems and components of the system should already have passed qualification and acceptance verification. For complex systems, the sum of the parts does not equal the whole due to interrelations between the parts. Therefore, even if all the subsystems have been approved, the whole system should be tested as well. This section discusses the different types of system testing along with system test process and technologies.

4.4.1 System Testing Types

System testing has a broad spectrum. It can be conducted for functional testing or nonfunctional testing such as performance testing, reliability testing, or

---

* In some test environments such as computer networks, it is possible to duplicate the actual environment.

compatibility testing. The nature of the system testing depends on the application domain. If the system is an aircraft or a missile, system test will involve some form of flight tests. If it is a vehicle, system test will involve some form of ground tests. If the system is a communication system, system test will involve some form of network testing. System test can be classified in two generic types based on the organization conducting the system test:

1. Development test and evaluation: The system test is conducted by the building organization. The tests are structured based on the system specification and engineering perspective. The system tests are implemented in special testing facilities and the system is instrumented to collect data in real time.
2. Operational test and evaluation: The system test is conducted by the customer. The tests are structured based on user's mission perspective and needs. Often, instrumentation equipment is not installed on the system under test. Therefore, it is more difficult to identify and track down the actual cause of any problems that arise during system testing.

Regardless of the type of test being conducted, system testing requires a systematic process as well. The following sections discuss the sequence of activities necessary for successful system testing.

4.4.2 System Testing Process

4.4.2.1 System Test Planning

A special kind of system test plan is prepared early in the program under the system engineering management plan. This plan is referred to as test and evaluation master plan (TEMP) or the integrated test and evaluation plan. This plan describes the management of the system testing and does not provide details of the system test procedures. The plan is updated during the design phase to guide the implementation of the system testing.

The detailed system test plan is a high-level verification plan based on high-level verification requirements. System test planning starts with defining the system test requirements that are derived from system specification. The next step is to synthesize the test requirements into a set of test tasks and select the optimum test strategy for the task. As the detailed test tasks are prepared, it is also necessary to define and plan for the supporting resources for the test tasks. The tools, test instrumentation, test facility location, and test environment configuration must be clearly planned during the planning of the test procedure. Test setup conditions, test case, and data collection should be stated for each test procedure. The complexity of the system, the planned use of the system, and the available budget allocated to testing determines the intensity of the system testing. A good test plan reduces testing cost by reducing redundant tests, guarantees customer satisfaction, and assures technical success by proving system performance and safety.

4.4.2.2 System Test Deployment and Management

System test deployment can proceed once the system test plan is complete and approved. Generally, system test deployment follows a similar process as V&V. Initially, there is a preparation phase where all resources are made available for the test. Then, system test is conducted and data is collected to analyze the performance of the system. The test engineer should monitor and control all testing closely and is usually charged with the responsibility of archiving all important test data.

The actual test procedure generally starts from simple test procedures and moves toward more complex ones. It may also start from safe and move toward more hazardous or from easy toward more difficult test procedures. Figure 4.4 illustrates a general outline for the system testing process.

Once the test results are analyzed, a decision is reached regarding the pass/fail of the test. This is documented in formal reports. If errors are detected or system does not fully satisfy the customer's missions, then the formal reports are fed back to the designers and developers for analysis and possible changes. All errors are not clearly identifiable and it is difficult to find poor performance errors due to the complex interrelations among system components. Therefore, earlier re-verification will be less costly. Later re-verification will cause more information to be changed.

4.4.2.3 Testing Technologies

Testing technologies vary depending on the test environment. High-fidelity real-time test environment requires the actual system to be run at the same rate and volume of its actual operational environment. In such cases, special instrumentation can be designed to support such a testing environment. The instrumentation can be embedded in the operational system and can consist of sensors, sensing connections, and the means to communicate and collect data.

In some cases such as space systems, it will not be always possible to create the actual test environment. For such cases, modeling and simulation provides reusable system frameworks and are effective tools to support system testing. The simulations can execute a sequence of test activities with different levels of fidelity.

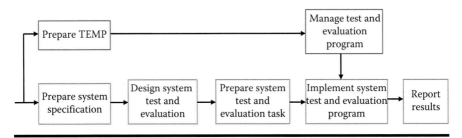

Figure 4.4   System testing process.

Specific to software systems, hardware-in-the-loop tests are necessary, which introduce hardware elements into the software test environment. Embedded systems that are time critical in nature should be tested in such a manner.

4.5 VV&T for Virtual Network Service and Next Generation Networks

Up to this point, we have discussed the generic VV&T processes. It is necessary to tailor the generic VV&T to the system of interest. A generic VV&T process model is developed and then applied to various industries including electronic/ avionics, automotive, food packaging, steel production, software-based large systems (Hoppe et al., 2007), real-time weapon systems, and data-acquisition systems (Lewis, 1992). In all cases, the generic VV&T methodology is adapted to the industrial and organizational characteristics of the company as well as the project types such as new product development and existing product upgrades. This section follows a similar approach and illustrate how the generic VV&T methodology is customized for the system of interest. For this, two systems are considered: The first system of interest is a virtual network service (VNS) that deploys intelligent network (IN) technology. The second system of interest is next generation networks (NGNs). The two examples highlight another important aspect: the adaption of VV&T for complex systems and for SoS. VNS is a complex system with dedicated type of network service (voice, data, etc.). NGN is an SoS, which integrates independent service types (voice, data, Internet, and media) into a single operational platform. The tailoring of VV&T for these systems illustrates how SoS testing is different compared to complex system testing.

4.5.1 Virtual Network Service

High-level architecture of a basic VNS is outlined as background information to supplement the VV&T process discussion for VNS. VNS systems are based on INs, where the service logic for each customer is external to the switching systems and is located in databases called service control points (SCPs). The benefit of IN is that new services can be rapidly introduced and customized according to a particular customer's needs through the service creation environment (SCE) and the service management system.

The VNS architecture is decomposed into two functionalities: transport and manage. Customers access the virtual private network through switches (SW) (action control points) using customer premises equipment (CPE) such as wireline and wireless phones. The signaling network transports the customer's query waiting at the switch to SCPs, which are databases that control the logic of the customer profile. Based on the response from the SCP, the transport network routes the customer's query to the terminating point. The network management subsystem focuses on fault detection, network configuration, network performance, and accounting. The service management subsystem (SMS) focuses on customer care and billing. This subsystem is integrated with SCE, which is interconnected to SCPs and implements customized service applications. Detailed information on VNS architecture can be found in Freeman (2004). Figure 4.5 illustrates the main components of VNS architecture.

The goal of VV&T is to identify and resolve problems in the overall end-to-end system before the service logic is deployed and put into service. The VNS is a complex service, which consists of many components, subsystems, and systems organized into an embedded structure. It is necessary to verify component-level end products, followed by verification of subsystems and the overall VNS. In this case, it is important to verify not only the physical network architecture and the network functionality but also the services that are embedded into the network. Therefore, the VV&T is customized to include service verification as well.

The VV&T of VNS starts with nodal testing. Nodal tests focus on verification of network components such as SCP, signaling network known as signal transfer points (STP), SW, transmission facilities, and CPE. Once all components are qualified, interface tests are conducted to verify the functional interfaces among different nodes such as interface between CPE and SW, SW and STP, and STP and SCP. Then, network testing is conducted to qualify the overall network; this is

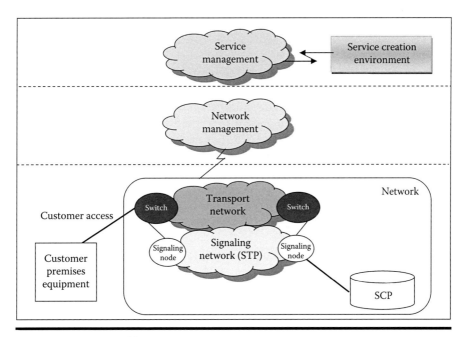

Figure 4.5   VNS architecture.

an end-to-end testing from originating to terminating station to verify that the call was routed according to the intended logic of the customer's profile. Once the network functionality has been verified, the end-to-end "subscriber" testing will validate that the service is provided as intended (service validation test); for example, proper records are generated at the network nodes to track end-user activity and these records are properly collected by the service management systems for billing. Operations readiness tests (OPR) are conducted to ensure that the processes, people, and organizations are ready to support the new service; these include sales, customer profile creation and downloading, customer care agents answering questions about mishandling of calls, billing, failures, etc. Each of the VNS subsystems, SMS, SCE, customer care, billing, and network management have to be tested for its corresponding system functionality (systems test) prior to the network, service, or operational readiness test. The end-to-end tests verify functional and nonfunctional elements of the VNS. Before the actual operation of VNS, acceptance tests are conducted as evidence for performance of the system; these tests are conducted by the end users (customer, network operations people, VNS administrators, sales force, etc.). Finally, system testing verifies VNS in its actual operating environment, and services are validated by the customer. Table 4.4 summarizes the customized VV&T for VNS.

The VV&T activities are classified into verification, testing, and validation in Figure 4.6. It is important to note that VV&T activities are conducted in parallel with integration. As components are integrated one at a time, related verification, testing, and validation activities have to be conducted to move on to the next level of integration.

VV&T planning for VNS starts with preparing the test requirements. Test network architecture requirements, test service management requirements, and the test physical location requirements are clearly defined. Based on these verification requirements, test cases (tasks) are identified and documented. Test procedure documentation should include information about the entry requirements, deliverables, status criteria, testing approach, and exit criteria. The plan should also provide the test implementation schedule and the responsible personnel for conducting the test cases. Once the tests are conducted, documentation of the test results is important. Table 4.5 illustrates a test case documentation sample for VNS.

4.5.2 Next Generation Network

During the twentieth century, the services were built in vertical architectures that were independent of each other having as a consequence service platforms dedicated to each type of service, that is, voice services networks, data networks (frame relay, asynchronous transfer mode, IP networks, etc.), private lines, and so on. With advent of IP and Internet, the NGN builds on a single IP platform to offer all types of service (voice, fax, data, video, and multimedia). Thus, the NGN evolves the

Table 4.4   Description of VV&T Activities for VNS

VV&T Types	Activity
Nodal testing	Network components' functionality, performance are tested
Interface testing	Interfaces between network components are tested (e.g., message flow)
Network testing	Overall network is tested from a to z (all network nodes are interconnected to establish the connection from a to z)
Operations readiness test	Ensure services are configured and functioning correctly in the network environment before customers start using the VNS (testing of the system behavior)
User acceptance test	Tests conducted to give confidence to end user regarding the operation and performance of the VNS in operation (different type of end users: customer, operation personnel, customer care agents, etc.)
End-to-end test	Testing the functional as well as nonfunctional elements of VNS
Operational (system) testing	VNS system testing in its actual operating environment; ensuring that the personnel and processes are available to deliver the intended service
Service validation test	Services are validated by the customer; this includes from customer ordering the system, to actual profile creation, customer usage, and rendering of billing statements

public switched telephone network (PSTN) and the public switched data network (PSDN) to create a multiservice network to integrate wireless and wireline services where the service functions are independent of the transport network. The result is a distributed SoS architecture. This section will focus on the VV&T environment for SoS by examining VV&T of NGN and highlight the challenges associated with verifying such systems. High-level architecture of a basic NGN system is outlined as background information to support the discussions. Figure 4.7 illustrates NGN architecture.

Customers access the NGN through end devices that comply with the user-to-network interfaces via desktops, laptops, wireless phones, PDA, etc. The access technologies accommodate a wide variety of wireline, wireless, cable, DSL, Ethernet technology, etc. The access network controls user profiles, user traffic, and also transports information across the access network in packet format. The access network interfaces with the core transport network to establish connectionless sessions from originating to terminating point; contrary to VNS, the IP-based multimedia has its intelligence distributed through the network layer, the network management layer,

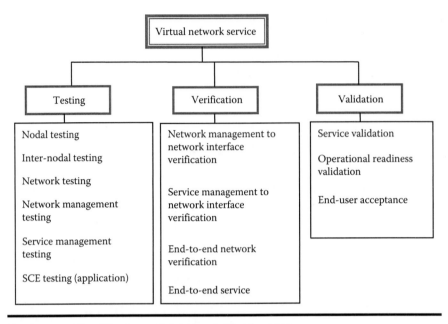

Figure 4.6 Classification of VV&T activities for VNS.

Table 4.5 Test Case Result Documentation

Service Verification	Test Case No: 56
Create new customer profile	Create a new customer profile in service management system
Expected outcome	Verify that the customer profile was successfully created
Test procedure	This will document the step-by-step procedures the tester must follow in order to create a customer profile
Test results	Success .... Failed ....
If failed enter problem encountered	
Tested by	
Supervised by	
Test Date and Time	

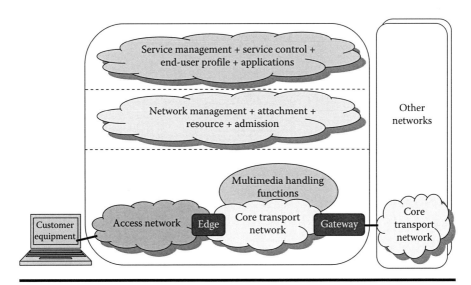

Figure 4.7   NGN architecture.

the services layer, and the application layer. The interfaces among all of the different subsystems determines what type of resources will be needed in the network (e.g., bandwidth and class of service), the quality of service, admission policies, authentication, etc. For instance, the network management controls admission, manages IP address space, transports user profile functions, and focuses on security, reliability, and quality. Similar to VNS, the NGN SMS is separated from the transport network. This subsystem is responsible for charging and billing customers, accounting, as well as customer support. The service management system supports open application interfaces that enable third-party service providers to create enhanced application services for NGN users through the application layer. The core transport network is integrated to other networks through gateways that provide capabilities to collaborate and interconnect with other networks such as Public Switched Telephone Networks/Public Switched Data Networks (PSTN/PSDN) and the Internet as well as other NGNs operated by other administrators. Detailed information on NGN architecture can be found in International Telecommunication Union—Telecommunications Standard Group (ITU-T) Rec. Y2011 (2004) and Boswarthick (2006).

The VV&T of NGN starts with nodal testing. Nodal tests focus on verification of network components such as CPE, access network nodes, and core transport network nodes. Once all components are qualified, interface tests are conducted to verify the functionality of interfaces such as edge node and gateway. Then, network testing is conducted to qualify the overall network. OPR are conducted to ensure SMS and network management subsystems are seamlessly configured in the network environment. The end-to-end tests verify functional and nonfunctional elements of the NGN. Before the actual operation of NGN,

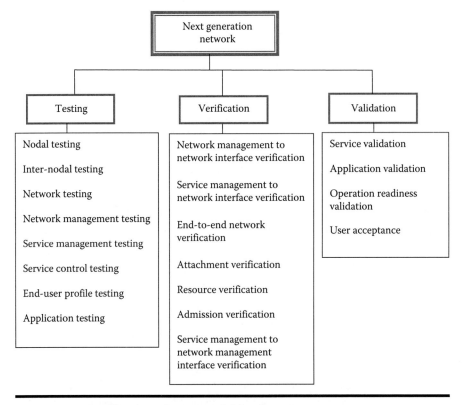

Figure 4.8   Classification of VV&T activities for NGN.

acceptance tests are conducted as evidence for the performance of the system. Finally, system testing verifies NGN in its actual operating environment and services and applications are validated by the customer. Figure 4.8 classifies the VV&T activities for NGN.

VV&T activities for NGN are similar to VV&T activities for VNS. However, NGN comprises a diverse set of control and management subsystems that will have to be tested during each of the V&V steps. This is an added challenge; because of the distributed nature of NGN, we need to link all the architectural components to ensure service interfaces compatibility and verification. As an example, during nodal testing, we do not only test for the node functionality but also have to check that the node is processing and assigning the resources as dictated by the control function in the network management layer and the service management layer. Therefore, the majority of the NGN verification activities focus on verification of the interfaces among NGN subsystems. It is necessary to confirm that service management, service control, end-user profile, and applications are seamlessly integrated into the NGN network as well as these functions seamlessly collaborate with network management functions. Besides, VV&T should assure that NGN can successfully interface with other networks. In such a complex testing environment, operational scenarios are usually not well defined or unexampled. This complicates generation of system test scenarios, thus it is not possible to test for all combinations of operational scenarios. Besides, the budget allocated to VV&T may limit the number of tests that can be conducted and also the time allocated to carry out the testing. Therefore, it is necessary to identify test selection criteria to detect severity of likely faults at an acceptable cost. These test selection criteria can then be translated into test case requirements. In most SoS testing environments, model-based testing in the form of simulations is utilized to support VV&T activities to reduce the cost.

Readers interested in more details on the NGN may find a good introduction to NGN in the IEEE Communications Magazine (October 2005) where there are several papers dedicated to NGN; see Carugi et al. (2005) and Sub-Lee and Knight (2005). NGNs are scheduled to be deployed sometime around the year 2012–2014; a nice summary of NGN plans may be found in Marco et al. (2005) and in Sub-Lee and Knight (2005).

4.6 Case Study: NASA Constellation Program

This section outlines the TEMP for the Constellation Systems Launch Vehicle (CSLV) that is being developed by NASA. TEMP provides the communication between project management team and the VV&T teams. It is an evolving document and is updated regularly to remain current and correct. The following TEMP outline is derived from the systems engineering management plan for the CSLV (v1, 2005) as an example to highlight the important sections and is not intended to cover all the details of an actual TEMP. Interested readers can find TEMP templates in Grady (2007) and Blanchard (2003).

4.6.1 CSLV Test and Evaluation Master Plan (Outline)

1 **Executive Summary:** This TEMP describes the overall activities for the test and evaluation of the CSLV project. It is written to provide a management plan to guide and assist the testing and evaluation activities. The scope of the TEMP applies to the CSLV. TEMP will be updated at each milestone review.

   1.1 **Mission Description:** The mission of the Constellation Program is to develop a new generation of spacecraft that will carry humans to the International Space Station, Moon, and then onward to Mars and beyond.

   1.2 **Constellation System Introduction:** Constellation system is comprised of spacecraft, launch vehicle, common support services, and destination surface segments. Launch vehicles deliver cargo and crew to Earth orbit as well as trans-lunar trajectories. Common support services include ground facilities and space-based assets that provide mission support to crew training, flight control, and communication. Destination surface segments include power systems, robotic systems, and resource utilization systems that support the crew members. The scope of the TEMP is spacecraft component of the Constellation System. It is referred to as CSLV.

1.3 CSLV System Description: Unlike space shuttle where both crew and cargo are launched together on the same rocket, CSLV project outlines two separate launch vehicles. CSLV is comprised of crew launch vehicle (CLV) and cargo launch vehicle (CaLV). CLV will provide safe and reliable transportation to the low earth orbit. CaLV will deliver 125 mT cargo delivery to low earth orbit. The CaLV will deliver cargo from low earth orbit to trans-lunar trajectories. Having two separate launch vehicles will allow for more specialized designs for different purposes.

2 Integrated Test Program Summary: This section describes the verification, testing, and validation activities for CSLV. VV&T of the launch abort system is outside the scope of the CSLV testing.

2.1 Integrated Test Program: The VV&T process will start with identifying the verification requirements. This includes development of verification requirements matrices for the specifications and the detailed verification requirements. The requirements will include qualification and acceptance verification requirements. Verification requirements will cover CLV, CLV subsystems, and CLV elements. These requirements will be documented in the CRADLE database. Each element will have its own V&V plan that will be linked to the related verification requirement in the CRADLE database. Following methods will be used for V&V: demonstration, inspection, analysis, test, simulation, and similarity. Following plans will accompany TEMP:

Formal development test plans
Qualification test plans
Acceptance test plans
Analysis plans
Master verification and validation plan
Test plans such as modal tests, structural tests, and separation tests
Software IV&V plans
Development test and evaluation plan
Operational test and evaluation plan

Each plan will contain the test description, test approach, test location, test conditions, and success criteria for each requirement.

2.2 Management: CSLV project organization is responsible for integrating and coordinating the CSLV project. Vehicle integration office (VIO) is responsible for controlling and developing verification requirements for CSLV. VIO is organized as shown in Figure 4.9.

Figure 4.9 Vehicle integration office organization.

3 **Test and Evaluation Resources:** This section identifies the resources needed to conduct VV&T.

   3.1 Test Sites and Instrumentation: The project requires a series of test facilities. Some of them are listed below. Additional test sites will be included in the TEMP as the program proceeds.

     A3 Test Stand: Test facility for testing the rocket engines and rocket stages that will help carry humans to the moon

     A1 Test Stand: Test facility for testing the rocket engines and its components

     B Complex Test Facility: Testing facility for testing the upper stage of the CLV and first stage of the CaLV

     Test Flight Pad: Test facility for flight tests for CLV and CaLV.

   3.2 Test Support Equipment: Some of the test support equipments for the CSLV Project include control systems, data acquisition and recording systems for test facilities, fire and gas detection systems, and liquid nitrogen pump system to support liquid nitrogen needs during testing.

   3.3 Simulations, Models, and Test Beds: Types of simulation and models are listed below:

     Simulations and models that replicate equipment operation or performance

     Environmental simulators

     Flight test simulators

     Crew module simulators

     Simulations and models that replicate combination of integrated equipments

4.7 Summary

This chapter focused on the VV&T activities for complex systems. Verification is the process of confirming that the system and its elements meet its requirements. It ensures that the system performs its intended function and meets performance requirements allocated to it. Validation is the process of determining whether the system functions in the manner its stakeholders/customers expect when used in its intended environment. A system can be verified and meet all its requirements, but this does not guarantee that the system is validated. System testing is a subset of V&V, which focuses on evaluating system operation in its actual environment. The VV&T process starts with planning and proceeds to implementation and management. Documentation of the results is important and is managed by databases. It is important that VV&T activities are intermingled with integration activities and that integration moves up to the next higher level once all integrated components are verified, tested, and validated. The generic VV&T activities are tailored to the system of interest. As examples, the tailored VV&T activities for VNS and NGN systems are discussed. As several independent networks are integrated in NGN to form SoS, challenges and constraints arise for the VV&T of NGN in terms of unexampled operational scenarios and limitations in test scenario generation. Finally, test planning is an important step for successful test implementation. TEMP outline is derived for the NASA's CSLV.

Authors

Ricardo L. Pineda holds PhD and MSc degrees from Lehigh University and a BSc degree from Universidad Nacional de Colombia. He has over 20 years of experience in systems engineering (SE) in different industries ranging from research and development at Bell Labs to chief technology officer at AT&T in Mexico. He was a distinguished member of the technical staff at Bell Labs where as a systems engineer he worked on requirements and the architecture of new technologies used in the AT&T network. He was the chief technology officer for the definition, development, implementation, and deployment of the AT&T network and services in Mexico. He was also practice director of Siemens Business Services for Latin America where he was the main consultant in systems implementations in Venezuela, Colombia, Ecuador, and Brazil. Dr. Pineda has extensive experience in academia; he was a professor at ITESM in Monterrey, Mexico, and at the "Universidad de Los Andes" in Colombia. His current research projects include PI for "Prognosis & Resilience Design for Complex SoS" with Raytheon-IDS, PI "SOS Global Attributes to Design Space Mapping" and "Technology Refreshment Assessment Model" for LMC-Aero, PI "Level of Repair Analysis Models" for Hamilton-Sundstrand, and PI for the "TMAC El Paso del Norte-Region" for the MEP program sponsored by NIST. Dr. Pineda was nominated to receive the "US-Hispanic Engineer of the

Year Award," received the "AT&T Architecture Award," and was honored with the "Baldwin Fellowship" and "Gotshall Fellowship" awards.

Currently, he is at the University of Texas at El Paso where he is the director of the Systems Engineering Program, the director of the Research Institute for Manufacturing and Engineering Systems, and the chair of the Industrial, Manufacturing and Systems Engineering Department. He is also a member of International Council on Systems Engineering, IEEE, and ISACA.

Nil Kilicay-Ergin is currently assistant professor of SE at the Penn State University's School of Graduate Professional Studies. Prior to joining Penn State University, she worked as a visiting research assistant professor within the Research Institute for Manufacturing and Engineering Systems at the University of Texas at El Paso where she taught for the SE graduate program and served on industry-funded research contracts, including research on prognostics design for system-of-systems (SoS) and research on SoS modeling. Dr. Ergin was also a postdoctoral fellow at the University of Missouri-Rolla where she conducted research on multiagent modeling techniques for SoS analysis.

Dr. Ergin earned her PhD in SE and MS in engineering management from the University of Missouri-Rolla (currently known as Missouri University of Science & Technology). She also holds a BS degree in environmental engineering from Istanbul Technical University, Turkey. She has been on the review committee of several conferences and journals including IEEE International Conference on System of Systems Engineering, ANNIE Conference Series, and the *International Journal of General Systems*. She is a member of International Council on Systems Engineering and the International Test and Evaluation Association.

References

Blanchard, B. 2003. *Systems Engineering Management*. Wiley: Hoboken, NJ.

Boswarthick, D. 2006. Technical Officer TISPAN, Global Standards, the Key Enabler for the Next Generation Network, *Expocomm 2006 Presentation*. Mexico City, Mexico.

Carugi, M., Hirschman, B., and A. Marita. 2005. Introduction to ITU-T NGN focus group release 1: Target environment, services and capabilities. *IEEE Communications Magazine*. 43(10): 42–48.

Constellation Systems Launch Vehicle (CSLV) Systems Engineering Management Plan (SEMP), version 1.0, May 2006, http://spacese.spacegrant.org/SEModules/Management/CxP-72018,%20CLV%20Sys%20Engr%20Mgmt%20Plan%20%28SEMP%29.pdf, last accessed August 2010.

Freeman, R.L. 2004. *Telecommunication Systems Engineering*. Fourth edition, Wiley: Hoboken, NJ.

Grady, J. 2007. *System Verification*. Academic Press Elsevier: Amsterdam, the Netherlands.

Hoppe, M., Engel, A., and S. Shachar. 2007. SysTest: Improving the verification, validation, and testing process—Assessing six industrial pilot projects. *Systems Engineering Journal*. 10(4): 323–347.

INCOSE Systems Engineering Handbook, version 3.1, INCOSE, 2006.

International Telecommunication Union–Telecommunication Standards Group (ITU-T) Rec. Y2011, *General Principles and General Reference Model for Next Generation Networks*, 2004.

Jain, R., Chandrasekaran, A., Elias, G., and R. Cloutier. 2008. Exploring the impact of systems architecture and systems requirements on systems integration complexity. *IEEE Systems Journal*. 2: 208–223.

Kaner, C., Falk, J., and H.Q. Nguyen. 1999. *Testing Computer Software*. John Wiley& Sons Inc.: Hoboken, NJ.

Lewis R. 1992. *Independent Verification and Validation*. Wiley-Interscience: Hoboken, NJ.

Levardy, V., Hoppe, M., and E. Honour. 2004. Verification, Validation & Testing Strategy and Planning Procedure. *Proceedings of the 14th International Symposium of INCOSE*. Toulouse, France.

NASA Systems Engineering Handbook, Revision 1, NASA/SP-2007-6105, NASA, 2007, http://education.ksc.nasa.gov/esmdspacegrant/Documents/NASA%20SP-2007-6105%20Rev%201%20Final%2031Dec2007.pdf, last accessed August 2010.

Sub-Lee, C. and D. Knight. 2005. Realization of the next generation network. *IEEE Communications Magazine*. 43(10): 34–41, October 2005.

Chapter 5

Assessing Technology Maturity as an Indicator of Systems Development Risk

Cynthia C. Forgie and Gerald W. Evans

Contents

Technological innovations are frequently the cornerstone of space exploration and critical decisions regarding which types of technologies to employ have a profound influence on the course of a program. Furthermore, innovative programs like those undertaken by the National Aeronautics and Space Administration frequently rely on technology forecasts to provide the input for research and development initiatives. While there is an extensive body of literature on the subject of project management, mitigation of risk attributed to technology development and systems integration (SI) issues is a frequently overlooked component of the program management process. Technology integration as part of a systems engineering life cycle needs an objective, quantitative assessment tool. This chapter discusses a framework for assessing technology integration at the system level. The chapter begins with a discussion of technology forecasting and some common techniques for assessing technology maturity. Then, a framework for addressing technology risk and integration issues at the system level is presented. This discussion is followed by a hypothetical example and concluding remarks.

5.1 Introduction

Technological innovations are frequently the cornerstone of new systems or program developments. For example, Constellation is a long-term strategic initiative managed by the National Aeronautics and Space Administration (NASA). The overarching goal of this program is the development of technologies needed to expand the science of space exploration and open the space frontier to Mars and beyond (GAO, 2006). Obviously, realization of a successful, human space program exploration program cannot be achieved without a considerable investment in technology. Furthermore, because program success is highly dependent on the timely development of several new technologies, program managers must determine the optimum portfolio of technologies necessary to achieve specific performance goals, while simultaneously managing cost and time resources.

Critical decisions regarding which types of technologies to employ have a profound influence on the course of a program. Pioneering programs like those undertaken by NASA frequently rely on technology forecasts to provide the input for research and development initiatives. However, no technological device springs directly from the mind of the innovator to widespread use. Normally, a technology evolves through a series of stages until it is sufficiently proven and ready for implementation.

Unfortunately, simply selecting and funding a technology is no guarantee that it will mature on time within budget and possess the required performance capabilities. Consequently, planning and managing strategic technology development programs such as the Constellation Program is an extremely complex task.

While there is an extensive body of literature on the subject of project management, mitigation of risk attributed to technology development and SI issues is a frequently overlooked component of the program management process. These issues become magnified when one considers the management of an entire portfolio of technologies and the multifaceted interrelationships inherent to a strategic systems development program. The purpose of this paper is to review the history of technology forecasting, discuss current techniques used for assessing technology maturity, and present a framework for evaluating readiness at the system level.

5.2 History of Technology Forecasting

Technology forecasting can be defined as a prediction of the future characteristics and timing of machines, materials, procedures, or techniques. Technology forecasting, in some form or another, has a long history. Early examples of forecasting include the Oracles of Delphi and Nostradamus. Ancient Egyptians foretold harvests from the level reached by the Nile in the flood season. In the seventeenth century, Sir William Petty suggested a basis for economic forecasts based on a 7 year business cycle, and today there are those that believe that the markings of the Woolly Worm can foretell the severity of the upcoming winter season.

Prior to the nineteenth century, the notion of technological change in society was relatively nonexistent. Consequently, the concept of technology forecasting is relatively new. The first published example of an attempt to anticipate technological changes appeared in a 1937 report published by the U.S. government entitled "Technological Trends and National Policy" (Martino, 1975, 648). After World War II, technology forecasting by both defense and civilian industries grew rapidly, and by the end of the 1960s, tools and techniques to perform technology forecasting had become commonplace in both the public and private sector. In 1972, the government formed a permanent office of technology assessment under the authority of the Technology Assessment Act. The purpose of this office is to provide objective analyses of major public policy issues related to scientific and technological change (92d Congress Public Law 92-484, 1972).

Two reasons for the development of more formal methods of technology forecasting are the increasing rate of technological advances and the growing emphasis on strategic planning. Technological advances are not free, and the increasing investment in industrial research and development are often critical for corporate growth and survival. Thus, major technological decisions could not be left to chance. Secondly, as corporations grow in size and complexity, the need for formal corporate planning necessitates the use of formal technology forecasting techniques.

Many techniques for technology forecasting have evolved over the last few decades and scholars have proposed a variety of ways to categorize these methodologies. The classification schema presented here describes eight categories of technology forecasting techniques: genius forecasting, environmental monitoring, consensus methods, trend extrapolation, simulation models, cross-impact matrix methods, scenarios, and stochastic methods.

Genius forecasting is based on a combination of intuition, insight, and luck. Gypsy fortune tellers and psychics are extreme examples of genius forecasting. Mainstream science generally ignores this type of forecasting because the implications are too difficult to accept and are also beyond mankind's current understanding of reality (Martino, 1975).

Environmental monitoring is based on the assumption that technological change is foreshadowed by changes in political, technical, economic, ecological, or social environment. Frequently, there are several precursor events that make it possible to forecast the eventual development of breakthrough technologies. For example, closing the Suez Canal necessitated changes in transportation routes for sea-going vessels, which eventually resulted in increased tanker size and capacity (Porter et al., 1980). Once a breakthrough event has been identified, it can be used to provide advance warning of the emergence of new technology. Although environmental monitoring lacks sophisticated analytical techniques, if used properly, monitoring helps focus forecasting efforts on problems that may be of strategic importance and plays a role in contingency planning (Jones and Twiss, 1978).

Consensus methods involve seeking opinions from one or more individuals and synthesizing these opinions into a final forecast. This type of methodology facilitates accumulations of two types of expertise. The first belongs to people with extensive knowledge about a specific subject area or topic. The second rests in general representatives of a group whose perceptions, attitudes, or actions influence the event to be forecast (Porter et al., 1980). Consensus methods range widely in complexity, cost, and time requirements. Some of the more well-known practices include the Delphi survey technique, nominal group method, and interview processes.

Trend extrapolation methods examine trends and cycles in historical data, and then use mathematical techniques to extrapolate to the future. Extrapolation methods assume that the forces responsible for creating the past will continue to operate in a similar fashion in the future. There are numerous techniques for forecasting trends and cycles. The most common involves various forms of weighted smoothing methods. A second type of method, known as decomposition, mathematically separates historical data into trend, seasonal, and random components. Autoregressive integrated moving average models such as adaptive filtering and Box-Jenkins analysis constitute a third class of mathematical model, while simple linear regression and curve fitting is a fourth class. The common feature of these methods is that historical data are the only information needed for producing a forecast (Porter et al., 1980).

Simulation models use a representation of a system to portray the dynamic behavior of that system as it varies over time. The model is an abstraction based on definable causal relationships exhibited by the original system. Simulation models can take on several forms including mechanical, mathematical, metaphorical, and gaming. An example of a mechanical model would be a cockpit mock-up used to train airline pilots. Common mathematical models include the S-curve, multiple regression analysis, Fisher–Pry analysis, Gompertz analysis, and Learning Curve techniques (Martino, 1975, 103). A metaphorical model could involve using the growth of a bacteria colony to describe human population growth. Game analogs involve the creation of an artificial environment in which players are assigned a specific role to perform. The number and diversity of simulation models is extensive and advances in computing technology have contributed to the increased popularity of simulation modeling.

Cross-impact analysis provides a logical framework within which to analyze the mutual influence of events and their interrelationships. It includes a group of analytical techniques that deal with the way in which impacts interact to produce higher-order impacts and how the occurrence of certain events affects the likelihood of other events. Examples of cross-impact analysis techniques used in technology forecasting include morphological matrices, impact analysis, content analysis, stakeholder's analysis, and patent analysis. The primary advantage of cross-impact analysis techniques are that they force decision makers to examine the interrelationships between system components and consider a systems view, rather than considering each system component as an individual entity (Jones and Twiss, 1978).

The scenario is a narrative sketch of a possible future state and offers a wide range of potential applications. For example, scenarios may be developed to forecast potential applications of a new technological concept, global population shifts, variations in global climatic conditions, or changes in consumer spending. Generally, scenarios are more powerful when contrasted with other scenarios. Therefore, a minimum of three scenarios—one optimistic, one pessimistic, and one most likely—are normally developed to describe a long-term prediction of the future. Because scenarios are applicable to a diverse set of issues, they are typically used to provoke thinking amongst decision makers and are considered a valuable planning tool (Jones and Twiss, 1978).

Stochastic techniques are those techniques that emphasize probabilistic properties in understanding and predicting future behaviors. Advances in computing technology have made it possible to create extremely complex models comprised of multiple subsystems. Decision trees, utility theory, Monte-Carlo methods, and Bayesian probability formulations are examples of stochastic techniques that have been applied to forecasting (Jones and Twiss, 1978).

Obviously, no single forecasting technique is suitable for every situation. Often, the combination of multiple forecasts derived from a variety of techniques yields more reliable and accurate forecasts. Combining forecasts provides a means of compensating for shortcomings in an individual forecasting technique. Therefore, by selecting complementary methods, deficiencies attributed to one technique can be offset by the advantages afforded by another technique.

Failure to properly mature new technologies invariably leads to cost and schedule overruns in major systems development programs (GAO, 1999, 3). Therefore, once a technology has been selected and included as part of a systems development initiative, some methodology for managing the development process and assessing readiness for SI becomes necessary. The following paragraphs discuss some of the current methodologies for assessing the maturity of a technology development project.

5.3 Techniques for Assessing Technology Maturity

5.3.1 Technology Readiness Levels

The technology readiness level (TRL) metric represents one of the earliest formal techniques developed for assessing the technology development process. TRLs are intended to describe increasing levels of technological maturity as a concept progresses from an initial idea to a fully tested and proven device. Readiness levels are measured along an ordinal scale ranging from one to nine. A TRL value of 1 represents the lowest state of readiness, such as a white paper. A TRL value of 9 would be assigned to a fully tested and operational system. An example of a TRL 9 technology would be the fuel cells used to provide electric power to the space shuttle.

The TRL concept was first used by NASA in the late 1980s as part of an overall risk assessment process. By the early 1990s, TRLs were routinely used within NASA to support technology maturity assessments and compare the maturity level of different technologies (Smith, 2005).

The current TRL model is based on the generic definitions set forth in a paper by John Mankins, the former Director of the Advanced Concepts Office at NASA Headquarters and is described in Table 5.1 (Mankins, 1995). Mankins defines TRLs as "a systematic metric or measurement system that supports assessments of the maturity of a particular technology and the consistent comparison of maturity between different types of technology." In addition to the TRL metric, Mankins proposed a second metric called the Research & Development Degree of Difficulty (R&D^3). The R&D^3 is an attempt to define the level of difficulty anticipated during the maturation of a particular technology and is intended as a complement to the existing TRL metric. The R&D^3 value is based on an ordinal scale where each project is assigned an R&D^3 value of I, II, III, IV, or V. An R&D^3-I suggests that the probability of success in a "normal" R&D effort is 99%, an R&D^3-II suggests a probability of success in a normal R&D effort is 90%, an R&D^3-III suggests an 80% success rate, R&D^3-IV suggests a 50% success rate, and an R&D^3-V suggests a 20% success rate (Mankins, 1998).

Table 5.1   Technology Readiness Levels—NASA Definitions

Technology Readiness Level	Description
1. Basic principles observed and reported	Lowest level of technology maturation. At this level, scientific research begins to be translated into applied research and development
2. Technology concept and/or application formulated	Once basic physical principles are observed, then at the next level of maturation, practical applications of those characteristics can be invented or identified. At this level, the application is still speculative: there is no experimental proof or detailed analysis to support the conjecture
3. Analytical and experimental critical function and/or characteristic proof of concept	Active research and development is initiated. This must include both analytical studies to set the technology into an appropriate context and laboratory-based studies to physically validate that the analytical predictions are correct. These studies and experiments should constitute "proof-of-concept"
4. Component and/or breadboard validation in laboratory environment	Basic technological elements must be integrated to establish that the "pieces" will work together to achieve concept-enabling levels of performance for a component and/or breadboard. This validation must be consistent with the requirements of potential system applications. The validation is relatively "low-fidelity" compared to the eventual system
5. Component and/or breadboard validation in relevant environment	The fidelity of the component and/or breadboard being tested has to increase significantly. Basic technological elements must be integrated with reasonably realistic supporting elements so that the total applications can be tested in a simulated or somewhat realistic environment
6. System/subsystem model or prototype demonstration in a relevant environment (ground or space)	A representative model or prototype system would be tested in a relevant environment. At this level, if the only relevant environment is space, then the model/prototype must be demonstrated in space

(*continued*)

Table 5.1 (continued)   Technology Readiness Levels—NASA Definitions

Technology Readiness Level	Description
7. System prototype demonstration in a space environment	Actual system prototype demonstration in a space environment. The prototype should be near or at the scale of the planned operational system and the demonstration must take place in space
8. Actual system completed and "flight qualified" through test and demonstration (ground or space)	The end of true system development for most technology elements. This might include integration of new technology into an existing system
9. Actual system "flight proven" through successful mission operations	The end of last "bug fixing" aspects

Source: Adapted from Mankins, J. C., Technology readiness levels: A white paper. NASA, Office of Space Access and Technology, Advanced Concepts Office, 1995.

In a 1999 study, the Government Accounting Office (GAO) reviewed 23 technology development projects with a wide range of readiness levels (TRL 2 to TRL 9) when they were included in product development programs. Programs with key technologies at readiness levels 6–8 at the time of program launch were capable of meeting cost, schedule, and performance requirements (GAO, 1999). For example, Ford managed its voice-activated control technology to a TRL 8 before introducing it on the 1999 Jaguar. Likewise, the Department of Defense (DoD) matured a revolutionary periscope technology to TRL 9 before it was included on the Virginia class attack submarine. However, DoD programs that accepted technologies at a readiness level of 5 or less experienced significant difficulties attributable to immature technologies (GAO, 1999). For example, the key technologies for the Army's brilliant antiarmor submunition were at TRLs 2 and 3 when weapon system development began. At these levels, DoD had a significant gap in technology maturity at the start of the program and the gap was not closed until well into the development program. Furthermore, problems with the technologies were the main contributors to the program's cost and schedule overruns (GAO, 1999). Based on this study, the GAO recommended that technologies should be a TRL 7 or higher before being included in a system design (GAO, 1999). The study concluded that the greater the number of immature technologies, the greater the probability that a system would fail to meet performance requirements.

The utility of the TRL concept has been recognized and adopted by several public and private sector organizations. In a 2001 memorandum, the DoD, Deputy

Undersecretary of Defense for Science and Technology mandated the use of TRLs in all new major systems development programs. Science and technology executives were directed to conduct a TRL assessment for critical technologies identified in major weapon systems programs prior to the start of engineering and manufacturing development. The memorandum also specifies that TRLs are the preferred approach and should be used to provide an overall system risk assessment for all new major programs (DoD 5000.2-R, 2001, C7.6.3).

The 2009 DoD Technology Readiness Assessment Deskbook contains recommended guidance on best practices, lessons learned, and expectations for systems developers. This publication identified TRLs as the appropriate methodology for assessing technology maturity and provided detailed instructions on the proper use of TRLs. The TRL definitions provided in the Deskbook are similar to those developed by NASA and are listed in Table 5.2 (*DoD Technology Readiness Assessment Deskbook*, 2009, Appendix C). An alternate set of TRLs intended specifically for assessing the maturity of software projects is also provided in the Deskbook.

5.3.2 TRL Calculator

Developed by Nolte for the U.S. Air Force, the TRL calculator simplifies the process of determining the appropriate TRL for a given technology and is applicable to both hardware and software programs. The calculator is a Microsoft Excel spreadsheet application that allows the user to answer a series of questions about a technology project. Once the questions have been answered, the calculator displays the TRL achieved in a graphic format (Nolte, 2003). Since the calculator was developed for the U.S. Air Force, it is based on the TRL descriptions provided by the DoD.

The TRL calculator provides a snapshot of technology maturity at a given point in time and provides a historical picture of the development process. This calculator is currently the only tool available to assist with the task of determining the appropriate TRL level for a given technology project (Graettinger et al., 2002). Unfortunately, the statistical validity of the calculator has not yet been demonstrated and is the subject of an ongoing effort by the Air Force Research Laboratory.

5.3.3 TDRA Tool

The technology development risk assessment (TDRA) tool is currently under development at the NASA Ames Research Center. The ultimate goal of the TDRA tool is to provide an estimate of the likelihood that a technology development project successfully meets its milestones. The TDRA tool represents technology maturity and evolution using the TRL scale. Development time and the probability that a technology will evolve up the TRL scale is calculated as a function of time and

Table 5.2   DoD Technology Readiness Levels

Technology Readiness Level	Description
1. Basic principles observed and reported	Lowest level of technology readiness. Scientific research begins to be translated into applied research and development. Example might include paper studies of a technology's basic properties
2. Technology concept and/or application formulated	Invention begins. Application is speculative and there is no detailed analysis to support the assumption. Examples limited to paper studies
3. Analytical and experimental critical function and/or characteristic proof of concept	Active research and development is initiated. Includes analytical studies and laboratory studies to physically validate analytical predictions. Examples include components that are not yet integrated or representative
4. Component and/or breadboard validation in laboratory environment	Basic technological components are integrated to establish that the pieces will work together. This is relatively "low fidelity." Examples include integration of "ad hoc" hardware in a laboratory
5. Component and/or breadboard validation in relevant environment	Fidelity of breadboard technology increases significantly. Basic technological components are integrated with reasonably realistic supporting elements so that the technology can be tested in a simulated environment
6. System/subsystem model or prototype demonstration in a relevant environment	Representative model or prototype system, which is well beyond the breadboard tested for TRL 5, is tested in a relevant environment. Examples include testing a prototype in a high-fidelity laboratory environment or in simulated environment
7. System prototype demonstration in an operational environment	Prototype near or at planned operational system. Represents a major step up from TRL 6, requiring the demonstration of an actual system prototype in an operational environment, such as in an aircraft, vehicle, or space. Examples include testing the prototype in a test bed aircraft

Table 5.2 (continued) DoD Technology Readiness Levels

Technology Readiness Level	Description
8. Actual system completed and qualified through test and demonstration	Technology has been proven to work in its final form and under expected conditions. In almost all cases, this TRL represents the end of true system development. Examples include developmental test and evaluation of the system in its intended weapon system to determine if it meets design specifications
9. Actual system proven through successful mission operations	Actual application of the technology in its final form and under mission conditions, such as those encountered in operational test and evaluation. In almost all cases, this is the end of the last "bug fixing" aspects of true system development

Source: DoD, *Technology Readiness Assessment (TRA) Deskbook,* Appendix C, https://acc.dau.mil/CommunityBrowser.aspx?id=18545 (accessed September 1, 2009).

budget for each technology sector. Calculations are performed using a continuous Markov-Chain Monte Carlo model. Baseline TRL transition probabilities are based on heritage data and expert opinions and may be updated to reflect funding levels, technical problems, and management philosophy.

Once completed, this tool will be used to add a development risk metric to the overall assessment of a technology project. It will provide a benchmark for measuring progress, identifying critical components, and providing direction for risk mitigation efforts. In some cases, the potential performance gains for a riskier technology may justify the additional risk. Regardless, consideration of the development risk in addition to the potential payoff leads to a more informed investment and better decision making (Mathias et al., 2006).

5.3.4 Technology Road Mapping and Investment Planning System

Clark and Olds (2005) developed the technology road mapping and investment planning system (TRIPS) process to simulate the technology maturation process. The TRIPS methodology focuses on modeling technology maturation using discrete-time Markov chains. A technology's TRL represents a state of a Markov chain and the TRLs are linked together with a series of transition probabilities. As illustrated in Figure 5.1, transition probabilities are derived from a triangular distribution that is a function of the dollar investment and the technology's R&D[3]. The nominal funding level represents the program manager's estimate of the cost

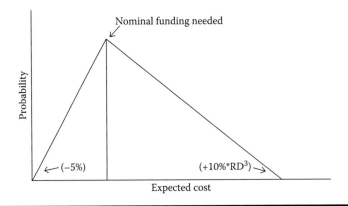

Figure 5.1 Triangular distribution of the cost to advance a technology from one TRL to the next TRL.

to develop a technology under "normal" circumstances. A separate triangular distribution is generated for each TRL. Preliminary investigations of the system have yielded promising results. However, TRIPS is still in the proof-of-concept stage and additional research is needed to move TRIPS forward.

5.3.5 Doctrine of Successive Refinement

Wheatcraft (2005) introduces a life cycle modeling technique called the Doctrine of Successive Refinement. The Doctrine of Successive Refinement is a spiral development process. The term "spiral" is used to describe the iterative process that is followed as the development of the technology is underway. An initial version of the system is developed and then repetitively modified based on customer feedback and evaluations. After each iteration around the spiral, a progressively more complete version of the technology project is created. This life cycle model is particularly applicable to complex projects when the requirements are not clearly understood or specified at the beginning of a project. Users may not know what they want until they see a prototype. Based on the prototype, requirements may be changed or modified and a new prototype generated. The user reevaluates the new prototype and makes additional recommendations. Based on the recommendations, a revised prototype is presented to the user for further evaluation. This cycle continues until a feasible concept or product emerges.

Wheatcraft suggests that taking a product from one TRL to another is analogous to completing one spiral under the Doctrine of Successive Refinement. As development progresses, requirements become more refined and as the technology progresses through TRLs, more and more operational capability is added. Rather than developing the completed product in one step, multiple spirals are completed.

5.3.6 *Technical Risk Assessment Framework*

Moon et al. (2005) developed a template for evaluating the technical feasibility, maturity, and overall risk of projects sponsored by the Australian DoD. This template referred to as the Technical Risk Assessment Framework, uses a combination of existing readiness scales to evaluate technical readiness. So, depending on the type of project, the user may choose the appropriate TRL scale. This template also provides a framework for applying risk mitigation strategies and identifying requirements for additional analysis.

The authors also incorporated the concept of systems readiness levels (SRLs) into the Technical Risk Assessment Framework. SRLs were originally proposed by the United Kingdom's Ministry of Defense (MoD) to address the technical risk at the system level and address SI issues (Sauser et al., 2006). The SRL concept represents a new system-focused approach for managing a system throughout the development life cycle.

5.4 Systems Readiness Levels

The SRL offers a new index for assessing system maturity and can supplement existing decision-making tools. The SRL approach addresses the technical risk and integration issues at the system level and offers an accurate and effective means of assessing system readiness and interoperability. For example, the system depicted in Figure 5.2 is comprised of two technology development projects. Each project has an associated TRL. Integration is the process of assembling the individual components into a single functioning system. The level of integration between the two technologies depicted in Figure 5.2 is referred to as the integration readiness level (IRL). Not only does each individual technology component need to be developed, the integration effort between technologies must also be completed to produce the desired system. Hence, the SRL is a function of both the maturity level of system components (TRL) and the IRL of the individual system components (Sauser et al., 2006).

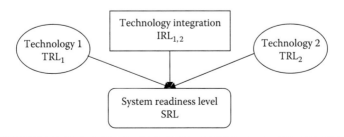

Figure 5.2 SRL conceptual framework.

Table 5.3 Description of Different IRLs

IRL	Description
1	The relationship between technologies has been identified
2	Low-level interactions between technologies and can be specified
3	The technologies are sufficiently compatible to support orderly and efficient interaction
4	There is sufficient detail in the quality and assurance of the integration between technologies
5	There is sufficient control between technologies necessary to establish, manage, and terminate the integration
6	Integrating technologies can accept, translate, and structure information for its intended application
7	Technology integration has been verified and validated
8	Actual integration completed and demonstrated in the expected operating environment
9	Integration is mission proven through successful mission operations

Source: Adapted from Sauser, B. et al., *Int. J. Ind. Syst. Eng.*, 3, 673, 2008a.

As the TRL is used to assess the risk associated with developing technologies, the IRL assesses the risk of integrating technologies. The IRL metric was initially proposed by Sauser et al. (2006) to provide a systematic measurement of the integration maturity of a developing technology with another technology. The IRL also supports consistent comparison of the maturity between integration points (i.e., TRLs). In addition to a reference as to the readiness of a technology for integration, IRLs also provide a direction for improving integration with other technologies (2008). The initial IRL metric was based on a seven-point scale. However, following additional research, the IRL scale has been revised into a nine-level scale with similar theoretical foundations as TRL (Gove et al., 2010). The IRL values listed in Table 5.3 were proposed by Gove et al. (2010) and are based on the revised nine-point scale.

5.4.1 SRL Formulation

Sauser et al. (2008a) proposed a methodology using normalized matrices of TRL and IRL indices to calculate a system's SRL value denoted as SRL_{System}. The computational approach for computing SRL_{System} can be summarized by the following five-step process: (1) establish a matrix comprised of the TRL values for each technology in the system; (2) establish a matrix comprised of the IRL values for each possible pair-wise integration between technologies; (3) normalize the TRL and IRL matrices to a [0,1] scale; (4) using the normalized TRL and IRL matrices, TRL′ and IRL′, respectively, compute the SRL matrix; and (5) compute the SRL index for the entire system, SRL_{System}. The SRL_{System} represents an index of maturity from 0 to 1 applied at the system level. These steps are described in further detail in the following paragraphs.

Step 1: Consider a system of n technology development projects. Each technology project has an associated TRL value ranging from 1 to 9. The TRL matrix defined in (5.1) consists of a single column containing the TRL values of each technology in the system, where TRL_i is the TRL for technology project i:

$$[TRL]_{n \times 1} = \begin{bmatrix} TRL_1 \\ TRL_2 \\ TRL_3 \\ \dots \\ TRL_n \end{bmatrix} \qquad (5.1)$$

Step 2: All possible pair-wise integrations between technologies are included in the IRL matrix, where $IRL_{i,j}$ is the IRL for technology projects i and j. The IRL matrix defined in (5.2) is of size $n \times n$. If there is no integration requirement between a specific set of technologies, then the IRL for that specific pair of technologies defaults to a value of 9. Because there is no requirement to integrate a technology with itself, the IRL values along the main diagonal of the matrix are assigned a default value of 9 ($IRL_{1,1} = 9$, $IRL_{2,2} = 9$, $IRL_{3,3} = 9$, …, $IRL_{n,n} = 9$). Also, based on the assumption that integration readiness between technologies is bidirectional, $IRL_{i,j} = IRL_{j,i}$:

$$[IRL]_{n \times n} = \begin{bmatrix} IRL_{1,1} & IRL_{1,2} & \dots & IRL_{1,n} \\ IRL_{2,1} & IRL_{2,2} & \dots & IRL_{2,n} \\ IRL_{3,1} & IRL_{3,2} & \dots & IRL_{3,n} \\ \dots & \dots & \dots & \dots \\ IRL_{n,1} & IRL_{n,2} & \dots & IRL_{n,n} \end{bmatrix} \qquad (5.2)$$

Step 3: Values used in the TRL and IRL matrices are normalized to a range of [0, 1]. Although the original values for both the TRL and IRL can be used, the normalized values are recommended to facilitate a more accurate comparison among different technologies. The normalized values are computed in (5.3). Let δ = the point to be normalized and δ' = the normalized value:

Table 5.4 Normalized Values

Original Value	Normalized Value
1	0.000
2	0.125
3	0.250
4	0.375
5	0.500
6	0.625
7	0.750
8	0.875
9	1.000

$$\delta' = \frac{\delta - \text{minimum possible value}}{\text{maximum possible value} - \text{minimum possible value}}$$

(5.3)

Let TRL′ and IRL′ represent the normalized TRL and IRL values, respectively. The normalized values for the nine-point [1,9] TRL and IRL scales are listed in Table 5.4.

Step 4: The SRL matrix is represented by the product of the normalized TRL′ and IRL′ matrices. It should be noted that the individual elements in the SRL matrix do not represent an index of maturity for each technology project. This methodology is intended as a system-wide assessment and cannot be applied to individual technology projects. Mathematically, for a system with n technologies, the SRL matrix can be computed as shown in (5.4):

$$[SRL]_{n\times1} = [IRL']_{n\times n} \times [TRL']_{n\times n}$$

$$\begin{bmatrix} SRL_1 \\ SRL_2 \\ SRL_3 \\ \vdots \\ SRL_4 \end{bmatrix} = \begin{bmatrix} IRL'_{1,1} & IRL'_{1,2} & \cdots & IRL'_{1,n} \\ IRL'_{2,1} & IRL'_{2,2} & \cdots & IRL'_{2,n} \\ IRL'_{3,1} & IRL'_{3,2} & \cdots & IRL'_{3,n} \\ \cdots & \cdots & \cdots & \cdots \\ IRL'_{n,1} & IRL'_{n,2} & \cdots & IRL'_{n,n} \end{bmatrix} \begin{bmatrix} TRL'_1 \\ TRL'_2 \\ TRL'_3 \\ \cdots \\ TRL'_n \end{bmatrix}$$

(5.4)

$$= \begin{bmatrix} (IRL'_{1,1}TRL'_1 + IRL'_{1,2}TRL'_2 + \cdots + IRL'_{1,n}TRL'_n) \\ (IRL'_{2,1}TRL'_1 + IRL'_{2,2}TRL'_2 + \cdots + IRL'_{2,n}TRL'_n) \\ (IRL'_{3,1}TRL'_1 + IRL'_{3,2}TRL'_2 + \cdots + IRL'_{3,n}TRL'_n) \\ \vdots \qquad \vdots \qquad \vdots \\ (IRL'_{n,1}TRL'_1 + IRL'_{n,2}TRL'_2 + \cdots + IRL'_{n,n}TRL'_n) \end{bmatrix}$$

Step 5: The SRL index for the complete system is computed as shown in (5.5):

$$SRL_{\text{System}} = \frac{\dfrac{SRL_1}{n} + \dfrac{SRL_2}{n} + \dfrac{SRL_3}{n} + \cdots + \dfrac{SRL_n}{n}}{n} = \frac{SRL_1 + SRL_2 + SRL_3 + \cdots + SRL_n}{n^2}$$

(5.5)

The computed SRL_{System} value represents a system level index of maturity ranging from 0 to 1. To be meaningful, the SRL must be correlated to the systems

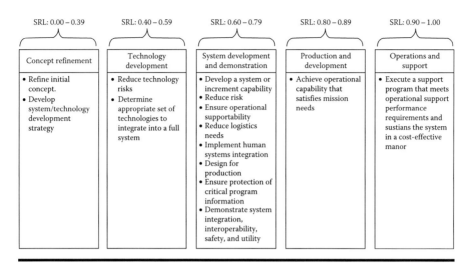

Figure 5.3   Description of the SRL index values.

engineering (SE) life cycle. The index values listed in Figure 5.3 were developed by Sauser et al. (2008b, 48) and correlate a system's current state of development in relation to the U.S. DoD life cycle management framework. The resultant SRL index provides an assessment of the overall system's status and a means of measuring project maturity. Because many systems cannot be verified in their operational environment until deployed, the mapping presented in Figure 5.3, also allows a system that has not reached full maturity to transition to the production phase. While the DoD phases of development are consistent with other life cycle models, practitioners may find it necessary to develop alternate mappings and correlations to meet specific organizational practices.

5.4.2 Example SRL Calculation

The following example uses a simple system to illustrate computation of the SRL value. The hypothetical system illustrated in Figure 5.4, is comprised of three technologies and three integrations. Technology project 1 is at TRL 6, technology project 2 is at TRL 8, and project 3 at TRL 9. The final system requires integration of the three technology projects, so there are three pair-wise combinations: projects 1 and 2, projects 2 and 3, and projects 1 and 3. The IRLs assigned to these combinations are as follows: projects 1 and 2 are at IRL 5, projects 2 and 3 are at IRL 7, and projects 1 and 3 are at IRL 6.

Step 1: The system is comprised of three technologies, so $n = 3$. The corresponding TRL matrix defined in (5.6) represents the current TRL values for projects 1, 2, and 3, where TRL_1 represents the TRL value for project 1, TRL_2 represents the TRL value for project 2, and TRL_3 represents the TRL value for project 3:

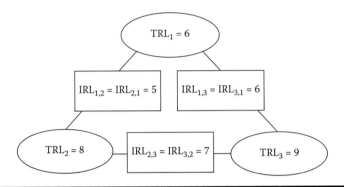

Figure 5.4 Example system.

$$[TRL] = \begin{bmatrix} 6 \\ 8 \\ 9 \end{bmatrix} \qquad (5.6)$$

Step 2: All potential pair-wise integrations between technology projects are included in the IRL matrix. Since $n = 3$, the IRL matrix defined in (5.7) is of size 3×3. Note that when $i = j$, the values on the main diagonal ($IRL_{1,1}$, $IRL_{2,2}$, and $IRL_{3,3}$) are assigned a value of 9. This assignment was based on the assumption that there was no requirement to integrate a technology with itself. Also, note that the matrix is symmetric ($IRL_{i,j} = IRL_{j,i}$):

$$[IRL] = \begin{bmatrix} 9 & 5 & 6 \\ 5 & 9 & 7 \\ 6 & 7 & 9 \end{bmatrix} \qquad (5.7)$$

Step 3: The values used in the TRL and IRL matrices are normalized to a range of [0, 1] from the original [1, 9] scale. The normalized values TRL′ and IRL′ are shown in (5.8) and (5.9), respectively:

$$[TRL'] = \begin{bmatrix} 0.625 \\ 0.875 \\ 1.000 \end{bmatrix} \qquad (5.8)$$

$$[IRL'] = \begin{bmatrix} 1.000 & 0.500 & 0.625 \\ 0.500 & 1.000 & 0.750 \\ 0.625 & 0.750 & 1.000 \end{bmatrix} \qquad (5.9)$$

Step 4: The SRL matrix is computed using the normalized matrices TRL′ and IRL′. This process is shown in (5.10):

$$[SRL] = \begin{bmatrix} SRL_1 \\ SRL_2 \\ SRL_3 \end{bmatrix} = \begin{bmatrix} (1.000)(0.625) + (0.500)(0.875) + (0.625)(1.000) \\ (0.500)(0.625) + (1.000)(0.875) + (0.750)(1.000) \\ (0.625)(0.625) + (0.750)(0.875) + (1.000)(1.000) \end{bmatrix} = \begin{bmatrix} 1.689 \\ 1.938 \\ 2.046 \end{bmatrix}$$

$$(5.10)$$

Step 5: The SRL index for the complete system is the average of the normalized SRL values and is computed in (5.11). Using the mapping presented in Figure 5.3, this system development would be in the system development and demonstration phase of its life cycle:

$$SRL_{System} = \frac{1.689 + 1.938 + 2.046}{3^2} = 0.630 \qquad (5.11)$$

5.5 Alternatives to TRLs and TRL-Based Methodologies

While there is considerable evidence to support the use of TRLs, there are also some difficulties associated with the TRL scale. The primary problems are (1) increased administrative overhead and training requirements, (2) inability to account for obsolescence and product aging, and (3) inability to address underlying variances between various technology sectors. Consequently, a variety of tools and techniques have been proposed to overcome these shortcomings. The following paragraphs discuss three promising alternative methodologies.

Mathias et al. (2006) developed a model that represents a range of technology development activities. This model deviates from the strict TRL classification scheme in favor of a six-stage development process. The key activities associated with each of the six stages along with mappings to the current TRL scale are presented in Table 5.5. Each stage has an associated underlying transition time represented by a normal distribution. Modeling parameters are based primarily on historical data, and Monte Carlo simulations are performed to generate the development time probability distributions used to assess the development risk for specific technology programs.

Smith (2004) proposed a methodology of assessing technology readiness based on Saaty's Analytic Hierarchy Process (AHP). Evaluation criteria or attributes are tailored to the particulars of a project. For example, readiness, environmental fidelity, product criticality, and obsolescence might be specified as the critical attributes. Each criterion is assigned a weight to reflect its overall impact on the project and the relative importance or status of each criterion is determined. For example, readiness

Table 5.5 Description of Key Activities Associated with the Six-Stage Methodology

Stage	Approximate TRL	Key Activities
1	1–2	Basic technology research and program initiation
2	2–3	Fundamental data acquisition for feasibility, determine requirements, and evaluating possible prototypes
3	3–5	Calculations, simulations, and subcomponent tests
4	5–6	Integration experiments, technology demonstrations, and prototype assembly
5	6–8	System/subsystem development, test, and refine prototypes
6	8–9	System test, deployment, and space operations

may be more important than obsolescence, or criticality may be significantly more important than readiness. The result of this process is a weighted score that represents the readiness level. Smith (2005) expanded this framework to include government off-the-shelf and open source software technology and products.

Valerdi and Kohl (2004) proposed an approach based on the Constructive Systems Engineering Cost Model (COSYSMO). COSYSMO is a parametric cost model that was developed to estimate the amount of systems engineering effort required for large-scale systems. It includes 18 drivers that capture the scope of a chosen system of interest. One of the drivers is the technology risk driver. This driver represents an effort to quantify technology risk on a five-point ordinal scale ranging from a "very low" to a "very high" level of technology risk. The COSYSMO working group is made up of over a dozen systems engineers with cost estimating experience and was responsible for the development of the technology risk driver definitions and rating scales.

5.6 Summary and Conclusions

Often, the decision to incorporate a new technology is critical to a program's success. Understanding and minimizing the risk of new technology, exploiting technology opportunities, and planning for technology maturity are key components of the SE process. This process is further complicated by the multifaceted interrelationships between individual technologies contained within a system. Unfortunately, mitigation of risk attributed to technology development is a frequently overlooked component of the SE process. In an era of economic uncertainty, large and expensive programs will be subjected to increased scrutiny and oversight. Therefore, it is crucial for the program manager to consider a program from a systems perspective and evaluate the impact of each distinct technology development on the overall program objectives.

TRLs can be instrumental in revealing a gap between a technology's maturity and the maturity demanded for successful inclusion in the overall system. TRLs provide a method of measuring a technology's maturity and can be used as an indicator of risk in the cost area, as well as in the technical performance domain. In the TRL model, as technology maturity increases, the overall development risk decreases (Moorhouse, 2001). However, due to the complex nature of today's systems, TRLs may not adequately assess SI issues or fully capture the risk involved when adopting a new technology (Valerdi and Kohl, 2004). Some researchers have suggested alternative methodologies such as a less restrictive measurement scale, Saaty's AHP, or the COSYSMO model to address this shortcoming. Another possible alternative, the SRL index, is an extension of the TRL-based methodology.

The SRL index provides a metric that can effectively determine system maturity, assess integration maturity, and provide a system-level readiness assessment. This index is a function of the individual TRLs in a system and their subsequent integration points, IRLs. The SRL index presented in this chapter defines the current state of development of a system in relation to the DoD's phases of development for the life cycle management framework. Practitioners may find it necessary to develop alternate mappings to meet the needs of specific organizational policies.

In summary, technology integration as part of a SE life cycle needs a quantitative assessment tool that can determine whether a collection of individual technology components can be integrated into a larger complex system. Furthermore, due to the size and complexity of many of today's systems, an objective and repeatable method is much needed (Graettinger et al., 2002, 16). The TRL model provides a framework for managing an individual technology development, and the SRL model provides a framework for managing a portfolio of technology developments throughout the SE life cycle.

Authors

Cynthia C. Forgie is an assistant professor of engineering at the University of Southern Indiana, Evansville, IN, United States. She received her BS, MS, and PhD degrees in industrial engineering from the University of Louisville. She also earned an MS in software engineering from Kansas State University. Dr. Forgie has served as a lecturer at the University of Louisville and Kansas State University. She has over 10 years of experience as an operations research specialist for the U.S. Army Operational Test Command and 5 years of experience as an industrial engineer with the U.S. Navy manufacturing base. In addition to simulation modeling and analysis, her research interests include decision analysis and program management.

Gerald W. Evans is a professor in the Department of Industrial Engineering at the University of Louisville. He has a BS degree in mathematics and MS and PhD degrees in industrial engineering, all from Purdue University. Before entering academia, he worked as an industrial engineer for Rock Island Arsenal and as a senior research engineer for General Motors Research Laboratories. Besides simulation modeling and analysis, his research interests include multi-objective optimization, decision analysis, and discrete optimization.

References

Clark, I.G. and J. Olds. 2005. The Technology road mapping and investment planning system (TRIPS). In Paper presented at the *AIAA 2005-4189, 41st AIAA/ASME/SAE/ ASEE Joint Propulsion Conference & Exhibit,* Tucson, AZ.

92d Congress, H.R. 10243. The Technology Assessment Act of 1972, Public Law 92-484, http://govinfo.library.unt.edu/ota/Ota_5/DATA/1972/9604.PDF (accessed February 10, 2009).

DoD. 2001. *Mandatory Procedures for Major Defense Acquisition Programs (MDAPS) and Major Automated Information System (MAIS) Acquisition Programs.* DoD 5000.2-R. http://mitre.org/work/sepo/toolkits/risk/references/files/DoD5000.2R_Jun01.pdf (accessed on February 1, 2009).

DoD. 2009. *Technology Readiness Assessment (TRA) Deskbook.* https://acc.dau.mil/ CommunityBrowser.aspx?id=18545 (accessed September 1, 2009).

GAO. 1999. *Best Practices: Better Management of Technology Can Improve Weapon System Outcomes.* GAO/NSIAD-99-162.

GAO. 2006. *U.S. Aerospace Industry Progress in Implementing Aerospace Commission Recommendations, and Remaining Challenges.* GAO-06-920. http://www.gao.gov/new. items/d06920.pdf (accessed February 26, 2008).

Gove, R., Sauser, B., and J. Ramirez-Marquez. 2010. Integration maturity metrics: Development of an integration readiness level. *Information Knowledge Systems Management,* 9(1): 17–46.

Graettinger, C.P., Garcia, S., Siviy, J., Schenk, R., and P. Van Syckle. 2002. *Using the Technology Readiness Levels Scale to Support Technology Management in the DoD's ATD/ STO Environments.* CMU/SEI-2002-SR-027, Pittsburgh, PA.

Jones, H. and B. Twiss. 1978. *Forecasting Technology for Planning Decisions.* Hong Kong: MacMillan Press Ltd.

Mankins, J.C. 1995. Technology readiness levels: A white paper. NASA, Office of Space Access and Technology, Advanced Concepts Office.

Mankins, J.C. 1998. Research and development degree of difficulty (R&D³): A white paper. NASA, Office of Space Access and Technology, Advanced Concepts Office.

Martino, J.P. 1975. *Technological Forecasting for Decision Making.* New York: Elsevier.

Mathias, D.L., Goodsell, A., and S. Go. 2006. Technology development risk assessment for space transportation systems. In Paper presented at the *Eighth International Conference on Probabilistic Safety Assessment and Management*, May 14–18, New Orleans, LA.

Moon, T., Smith, J., and S. Cook. 2005. Technology readiness and technical risk assessment for the Australian Defence Organisation. In Paper presented at the *Systems Engineering, Test & Evaluation Conference*, November 7–9, Brisbane, Australia.

Moorhouse, D.J. 2001. Detailed definitions and guidance for application of technology readiness levels. *Journal of Aircraft*, 39(1): 190–192.

Nolte, W.L. 2003. Technology readiness level calculator, Air Force Research Laboratory. In Paper presented at the *NDIA Systems Engineering Conference*, October 20–23, San Diego, CA.

Porter, A.L., Rossini, F.A., and S.R. Carpenter. 1980. *A Guidebook for Technology Assessment and Impact Analysis*. New York: North Holland Inc.

Sauser, B., Verma, D., Ramirez-Marquez, J., and R. Gove. 2006. From TRL to SRL: The concept of systems readiness levels. In Paper presented at the *Conference on Systems Engineering Research*, April 7–8, Los Angeles, CA.

Sauser, B., Ramirez-Marquez, J., Henry, D., and D. Dimariz. 2008a. A system maturity index for the systems engineering life cycle. *International Journal of Industrial and Systems Engineering*, 3(6): 673–691.

Sauser, B., Ramirez-Marquez, J.E., Magnaye, R., and W. Tan. 2008b. A systems approach to expanding the technology readiness level within defense acquisition. *International Journal of Defense Acquisition Management*, 1: 39–58.

Smith, J. 2004. ImpACT: An alternative to technology readiness levels for commercial-off-the-shelf (COTS) software. In Paper presented at the *Third International Conference on COTS-Based Software Systems*, February, Redondo Beach, CA.

Smith, J. 2005. An alternative to technology readiness levels for non-developmental item (NDI) software. In Paper presented at the *38th Annual Hawaii International Conference on System Sciences*, January 3–6, Big Island, HI.

Valerdi, R. and R. Kohl. 2004. An approach to technology risk management. In Paper presented at the *First Annual MIT Engineering Systems Division Symposium*, March, Boston, MA. http://web.mit.edu/rvalerdi/www/TRL%20paper%20ESD%20Valerdi%20Kohl.pdf (accessed December 14, 2007).

Wheatcraft, L. 2005. Developing requirements for technology-driven products. In *Compliance Automation Inc., International Council of Systems Engineering*. http://www.complianceautomation.com/papers/Wheatcraft_INCOSE_Techrqmts_032505.pdf (accessed November 20, 2007).

Chapter 6

Application of Business Process Reengineering to Space Center Workforce Planning

Gary P. Moynihan

Contents

Business process reengineering (BPR) infers a basic restructuring of essential business functions and processes, not merely their modification, enhancement, or improvement. The John F. Kennedy Space Center (KSC) workforce is undergoing such a fundamental restructuring process. This process involves organizational change, and a transition from a Space Shuttle operations orientation to a focus on development in support of the Constellation program. An overall plan for this process was developed, including consideration of KSC workforce skills and a review of operations across the KSC organization. Methods of data collection were developed in order to support these reviews. Fundamental to the development of this plan was the utilization of BPR techniques. This case study offers a potential framework of the change process that may be applied for further restructuring at KSC, or in the restructuring of other organizations.

6.1 Introduction

The National Aeronautics and Space Administration (NASA) has been undergoing a fundamental restructuring for over a decade. The diminishing budgets of the 1990s initiated this process. In 1993, the agency had approximately 25,000 civil servants in its Headquarters and Field Centers (NASA, 1997a). Through measures such as separation incentives, hiring freezes, normal attrition, and aggressive outplacement, NASA had fewer than 18,000 civil servants by the year 2000. This workforce transformation encompasses more than merely a reduction in numbers. During the late 1990s, the agency's strategic planning identified a transition in focus to research and development, rather than Space Shuttle operations. To accomplish this requires "reducing the infrastructure at Headquarters and the Centers; transferring operational activities to commercial contractors as appropriate, and focusing internal efforts on technology development" (NASA, 1997a). This planned transition took on new significance with NASA's 2004 report, "The Vision for Space Exploration," which outlined the agency's Constellation program to return to the Moon and Mars.

The John F. Kennedy Space Center (KSC) is NASA's primary launch site, and is the center of excellence for launch and payload processing systems. KSC's main focus relates to the remaining missions of the Space Shuttle in completion of the International Space Station (ISS). KSC is responsible for the preparation, launch, and landing of the Shuttle orbiter, crews, and payloads, as well as the subsequent recovery of the solid rocket boosters (NASA, 2008a). It is also the primary center for NASA's Launch Services program, which is responsible for launching satellites and robotic space missions on expendable launch vehicles (Cabana, 2009).

Consistent with NASA's changing strategic direction, the KSC Implementation Plan and Roadmap were developed in 1998. All subsequent Center workforce planning has used this as a foundation. The KSC Implementation Plan and Roadmap identify the Center's primary goals as it began transition from an operational work environment to one more focused on development. The primary goals are (Bridges, 1998)

"1. Assure that sound, safe, and efficient practices and processes are in place for privatized/commercialized launch site processing.
2. Increase the use of KSC's operations expertise to contribute to the design and development of new payloads and launch vehicles.
3. Utilize KSC's operations expertise in partnership with other entities (Centers, industry, academia) to develop new technologies for future space initiatives.
4. Continually enhance core capabilities (people, facilities, equipment, and systems) to meet Agency objectives, and customer needs for faster, better, cheaper development and operations of space systems."

With regard to the fourth goal, a zero-based review (ZBR) of the KSC workforce was conducted in 1997. A ZBR implies analyzing the need for each organizational function "from the ground up, since it starts with the assumption that the activity does not exist" (Knight and Wilmott, 2000). The ZBR identified yearly staffing levels though the fiscal year (FY) 2000, consistent with the agency strategic plan. However, subsequent concern was expressed regarding the identification and correction of skill mix imbalances (NASA, 1997b, 2006). The objective of this research was to develop a plan for an independent assessment of the Center's current work activities in order to support the decisions necessary to transition KSC toward a development work environment. The intent of the plan was to ensure that all employees are engaged in work linked with, and supportive of, the KSC Roadmap. Further, the plan would lead to the identification of skill gaps between present and future states, for retraining/redeployment efforts. This last issue remains a continuing concern as Center efforts shift to support the Constellation program (Gerstenmaier and Gilbrech, 2008).

6.2 Theoretical Framework

The concept of systems engineering entails a logical sequence of events which converts a set of requirements into a complete system description that fulfills the objective in an optimum manner (Skytte, 1994). It assures that all aspects of the project have been considered and integrated into the design solution. A variety of structured approaches to systems engineering have been defined over the past three decades (e.g., Blanchard and Fabrycky, 1998). Fundamental to the application of good systems engineering principles is an understanding of how critical factors relate to the overall design.

Consistent with these systems engineering principles, business process reengineering (BPR) was originated by Hammer and Champy (1993). They define reengineering as "the fundamental rethinking and radical redesign of business processes to achieve dramatic improvements in critical contemporary measures of performance, such as cost, quality, service, and speed." BPR infers a basic restructuring of essential business functions and processes, and not merely their modification, enhancement, or improvement.

As the name implies, BPR is based on the concept of analyzing processes. A process is "a collection of activities that takes one or more kinds of input and creates an output that is of value to the customer" (Hammer and Champy, 2006). According to classical BPR techniques (e.g., Malhotra, 1998), processes are composed of three primary types of activities: value-adding activities (i.e., activities important to the customer); transport activities (which move work across organizational boundaries); and control activities (which audit or control the other types of activities). Normally, a process may wind its way through many boundaries and controls within an organization. Each organizational boundary creates a transport activity and at least one control. Therefore, the more organizational boundaries that a process must transverse, the more non-value-adding activities that are present within the processes. It is this workflow path that BPR seeks to analyze and make more efficient.

Further, BPR focuses on strategic, value-added business processes, and the associated systems, policies, and organizational structures that support them. Strategic processes are those that are of essential importance to a company's business objectives. Value-added processes "are processes that are essential to a customer's wants and needs and that a customer is willing to pay for; they deliver or produce something that he or she cares about as part of the product or service offered" (Maganelli and Klein, 1994). Certainly, in order to achieve the maximum return on investment in a reengineering project, it is logical to begin by focusing on the most important processes in the organization. According to the literature, most corporations can be functionally decomposed into 12 to 24 processes, and usually no more than 6 of these are both strategic and value-adding (Khosrowpour, 2006).

Reengineering attempts to optimize the workflow and productivity in an organization. This optimization is measured in terms of performance indicators. For business corporations, these performance indicators may include reduction in cost, and increases in revenue and profitability. An important component of BPR is the deliberate and explicit mapping of these performance indicators with the processes to which they apply. Improving process workflow via BPR is merely the means to the goal of improving organizational performance as measured by these performance indicators.

Competitive pressures, and the associated desire to provide increased levels of value both to existing and future customers, initially led many companies to utilize this methodology. Successful BPR case studies, cited in the literature of the mid-1990s, included McDonnell Douglas (McCloud, 1993), BellSouth (Brittain, 1994), and Corning (Bambarger, 1994). Overuse and misuse of the concept resulted in a number of critiques (e.g., Knight and Wilmott, 2000). During the past 10 years, a more balanced view of BPR has emerged. Consideration of business processes has become an accepted starting point for organizational analysis and redesign, but is implemented in a less extreme manner than was originally proposed by Hammer and Champy (1993).

Although BPR was originally developed for the commercial sector, its applicability to government organizations was soon recognized. White (2007) notes that in 1994 the U.S. General Accounting Office (GAO) convened a symposium on BPR and the public sector. Based on the findings of this symposium, the GAO issued a guide for the application of BPR for federal agencies (U.S. GAO, 1997). Osborne and Hutchinson (2004) document several subsequent governmental BPR case studies:

1. The U.S. Department of Defense was able to reduce medical inventories by two-thirds.
2. The U.S. Postal Service reduced the time required for its formal approval process.
3. The U.S. Social Security Administration reduced the time required to determine disability eligibility from 155 days to 40 days.

Both McNulty (2004) and Khosrowpour (2006) discuss BPR initiatives by the British National Health Service. These successful public sector implementations provided precedents for the consideration of applying BPR principles to KSC workforce planning.

6.3 Application of BPR to KSC Requirements

Although the ZBR was completed in 1997, changing center resource requirements exceeded the scope of the original ZBR. In addition, KSC's core capabilities were not completely understood at that time. As a result, effective workforce management and development were limited in meeting the center's future requirements.

The Workforce Assessment and Validation Exercise (WAVE) was initiated as a follow-on to the ZBR, in order to conduct an independent assessment of the Center's current and future work activities and skills. This was required to support senior management's decisions toward transitioning from an operational work environment to one of development. Further, WAVE would identify skill gaps between these present and future states, and make recommendations for improved skills utilization. These recommendations would include retraining, redeployment, and reorganization.

This BPR project was conducted as part of the more comprehensive WAVE initiative. By utilizing the techniques and principles of BPR, the following plan was developed to meet these objectives. Consistent with other BPR approaches, the plan had four primary phases:

1. Organization and preparation
2. Identification of existing (or as-is) baseline
3. Determination of specific areas for transformation
4. Transformation of target areas into objective baseline

The complete BPR plan was both extensive and detailed, and is presented in Figure 6.1 as a work breakdown structure in order to provide a more concise presentation format. The plan is highly integrated. Relationships between specific steps are indicated. The individual sections of the plan are presented in detail in the following sections of this chapter.

6.3.1 Organization and Preparation (Phase A)

According to Maganelli and Klein (1994), the need for reengineering is usually recognized as a result of change, such as a market change, a technology change, or an environmental change. In this case, KSC's transition to a development orientation triggered the recognition for reengineering. A more immediate concern was the identification of skill gaps between the present and future states. A simple algorithmic model of the reengineering process may be considered to be

As-is baseline + benchmarking + goals/objectives = to-be baseline

The center's goals and objectives were originally established and documented in the KSC Roadmap (Bridges, 1998). Benchmarking is conducted as an ongoing process by a variety of groups at KSC. The benchmarking of other organizations provided ideas and insights for the modification of the as-is (or existing) baseline to attain the stated goals and objectives (which yields the to-be or objective baseline).

1.0 Organization and Preparation (Phase A)
- 1.1 Recognize need for restructuring
- 1.2 Establish executive consensus
  - 1.2.1 Secure support for project
  - 1.2.2 Identify scope of project and issues to be addressed
  - 1.2.3 Identify stakeholders in status quo
  - 1.2.4 Set goals and priorities for project
    - Includes prioritizing KSC roadmap goals and objectives
    - Linkage to business objectives and agreements (BOAs)
    - This prioritization should consider critical success factors for KSC
  - 1.2.5 Establish project charter
  - 1.2.6 Establish strategic plan/schedule
- 1.3 Organize reengineering team
  - 1.3.1 Define the approach
    - Functional decomposition (major functions and supporting activities)
    - Linkage to supporting matrices
  - 1.3.2 Identify team members and responsibilities
    - Core team
    - Functional teams
  - 1.3.3 Convey management expectations
  - 1.3.4 Allocation of sufficient resources
    - Manpower
    - Time
    - Budget
    - Technology
    - Training and consultants
    - Other considerations
- 1.4 Plan for change management
  - 1.4.1 Plan for means of communications with employees
  - 1.4.2 Obtain buy-in by current stakeholders
  - 1.4.3 Establish detailed plan/schedule for next phase

2.0 Identification (of existing or as-is baseline)
- 2.1 Identify primary functions/categories
- 2.2 Identify supporting activities and steps
  - Utilize ISO 9000 process flows as initial baseline (ref. Business World on KSC homepage)
  - Individual blocks within the process flows would be considered as steps
  - Each function, activity and step would be assigned a unique numerical identifier for use in the supporting matrices (see 2.4)
  - Linkage of skills to specific activities/steps
    - Utilize GPES system
    - Part of data validation by supervisor
  - 2.4.6 Additional data requested
    - Recommend that this data call be postponed until 5.3
    - If required at this time then data call includes:
      - Vision for the future
      - Ideas for reorganization
      - Willingness to reorganize
- 2.5 Roughcut skills gap determination
  - 2.5.1 Utilize employee's inputs to Glenn Software as existing skills baseline
  - 2.5.2 Modify Glenn Software to provide interface for management input

Figure 6.1   Major tasks in workforce review.

(continued)

- Provides input on perceived future skills requirements
- Quantity needed (FTE) per skill for specific FY
- Provides objective skills baseline
- May include validation loop by higher management

    2.5.3 Skills gap analysis
- Compare skills baselines and generate report of deltas
- Utilize Excel/Access Software
- Note: The objective skills file can not be used for subsequent BPR analyses. Further efforts are necessary.

3.0 Organization and Preparation (Phase B)
    3.1 Establish executive consensus
        3.1.1 Secure support for project
        3.1.2 Identify scope of project and issues to be addressed
        3.1.3 Identify stakeholders in status quo
        3.1.4 Set goals and priorities for project
- Includes prioritizing KSC roadmap goals and objectives
- Linkage to business objectives and agreements (BOAs)
- This prioritization should consider critical success factors for KSC

        3.1.5 Establish project charter
        3.1.6 Establish strategic plan/schedule
    3.2 Organize reengineering team
        3.2.1 Identify team members and responsibilities
- Continue with existing core and functional teams
- Modify team composition as necessary

        3.2.2 Convey management expectations
        3.2.3 Allocation of sufficient resources
- Manpower
- Time
- Budget

    2.1 Interview KSC personnel
- Confirm functional decomposition (Go to 2.4)
- Revise functional decomposition (Return to 2.2)

    2.2 Develop supporting matrices (X to Y)
        2.4.1 Activity/Steps to Function
- Matrix constructed as Part of 2.2

        2.4.2 Roles and missions
- Rows along Y axis: function, activity, step
- Columns along X axis: organization responsible. KSC goal/objective, external customer, internal customer, external supplier, internal supplier, performance criteria, executive priority
- Organizational responsibility:
  - ISO 9000 process flows indicate initial view of responsible organizations
  - Need to identify type of involvement on matrix, e.g., "responsible for," "provides information," "receives notification"
  - If necessary, this could be broken down to individual jobs within the organization, and their relation to activities and steps
- Use KSC roadmap for goals and objectives
- Performance indicators on KSC webpage (Under Business World Directory) already mapped to goals and objectives
- Variety of performance indicators identified; question of which are considered critical success factors to KSC
- Executive priorities set in 1.2.4

Figure 6.1 (continued)

2.4.3 Time matrix
- Rows along Y axis: activity usually, may occasionally be step
- Columns along X axis: frequency, volume, duration, and activity type (value adding, control, other)
- Matrix could be used to derive $ values, if necessary, by combining with other matrices and CFO databases
- Time matrix only required if Process Modeling software used in 5.2

2.4.4 Organizational charts (existing)
- Employees to organization
- Also identifies organizational resources (e.g., headcount)

2.4.5 Skills assessment
- Employees to skills
- Completed by individual KSC employees, rather than management
- Use Glenn Software and approach as basis
- Skills inventory must not only identify existing job-related skills But those that will be needed in the future (i.e., when KSC achieves a development orientation)
- Technology
- BPR Software
- Process Modeling Software
- Training and consultants

3.2.4 Plan for change management
- Plan means of communications with employees
- Obtain buy-in by current stakeholders
- Establish detailed plans/schedule for next phase

4.0 Determination (of specific areas for transformation)
- Need to clearly identify specific areas for analysis and restructuring
- Cannot restructure everything at once. According to BPR literature, this is a prescription for project failure

4.1 Identify set of new customers and their requirements
- Survey directorates/project offices
- Add new activities or modify existing activities in order to support new customers

4.2 Identify set of existing customers whose method of support will change
- Implied linkage (external and internal suppliers/internal customer/external customer)
- Modify activities to reflect change in method of support upon shift to development

4.3 Highlight any activities in this linkage that will not support 4.1 or 4.2

4.4 Focus on "core capabilities"
- Defined in literature as those that are both strategic and value-added
- Considered the most vital in the organization
- Linked to KSC core competencies
- At KSC, activities that directly affect stated roadmap goals and objectives (1.2.4)

4.5 Schedule efforts in next phase
  4.5.1 Near-term opportunities (over next 6 months)
  - Core process activities used currently and in the future upon transition
  - Sequenced by executive priority (1.2.4) as well as probability of success
  - Provide early demonstrated benefits of project
  4.5.2 Longer-term opportunities
  - Focus on core process activities, then on other high-value activities
  - Ignore any activities highlighted in 4.3
  - Sequence according to executive priority (1.2.4)
  - Develop phased transformation plan
    - Each phase addresses 1 or more "activity clusters" (i.e., linked activities)
    - Linkage implied by chain of suppliers and customers (see 4.2)

Figure 6.1 (continued)

- Avoids creating "islands of efficiency" among process bottlenecks
- Better leverages project efforts

5.0 Transformation (of target areas into objective baseline)
    5.1 Analysis of scheduled activities (4.5)
        5.1.1 Reexamine linkages to improve performance
- Steps within activity
  - Resequence as necessary
  - Identify any change in organizational responsibility
- Activities within clusters
  - Establish better coordination between activities
- Identify redundancies

        5.1.2 Define alternatives
- Eliminate redundancies
- Simplify activities/reduce steps
- Relocate and retime controls (2.4.3)
- Methods improvement
- Application of technology

    5.2 Evaluation of alternatives
        5.2.1 Evaluate each alternative based on resulting change in performance indicator (2.4.2)
        5.2.2 Recommend use of process modeling software
- Input original:
  - Activity flows (2.2)
  - Time matrix values (2.4.3)
  - Organizational resources based on organizational responsibility (2.4.4)
- Simulate existing baseline
- Modify activity flow and supporting matrices to reflect alternative.
- Simulate alternative activity flow to determine:
  - Delta of time requirements
  - Delta of manpower requirements

        5.2.3 Review of alternatives by blue ribbon panel
        5.2.4 Accept/revise alternatives (return to 5.1.2) based on delta of performance indicator
- Particularly delta of critical success factors (2.4.2)

    5.3 Comparison of accepted alternatives to original baseline
        5.3.1 Determination of impact on workforce
- Compare accepted alternate activity flow (5.2.4) to original (2.2)
- Update subordinate matrices:
  - Roles and missions (2.4.2)
  - Time matrix (2.4.3)
- Determine delta of skills (2.4.5) and headcount (2.4.5) for each alternative
- Consideration of core competencies/core capabilities provides priority

        5.3.2 Organizational change
- Specify management structure
  - Obtain KSC personnel's vision of the future
    - Ideas for reorganization
    - Willingness to reorganize
  - Benchmark other organizations
  - Possible alternative structures for consideration
  - Evaluation of alternatives and recommendation
- Redraw organizational boundaries
  - Use recommended management structure as framework
  - Organize around functions/activities

Figure 6.1 (continued)

- - ■ Consolidate resources
    - ■ Avoid duplication
    - ■ Streamline processing
  - ■ Specify retraining/redeployment
    - ■ Identify specific available candidates
      - ■ Employees associated, with activities no longer required in objective environment (4.3)
    - ■ Review candidates' skills in terms of type and level
    - ■ Compare to organizational skill needs (5.3.1)
    - ■ Plan Redeployment if match
    - ■ Plan retraining as necessary to provide match
      - ■ Weighted change in acquisition of skill level
5.4 Implementation
  - ■ Determine method for system validation
    - ■ Parallel test versus pilot test
  - ■ Determine Methods for conversion and transition
    - ■ Systems
    - ■ Procedures
    - ■ Documentation
    - ■ Data conversion
  - ■ Develop time-phased deployment plan
  - ■ Pilot new process
  - ■ Refine and transition
  - ■ Continuous improvement

Figure 6.1 (continued)

Initial efforts were focused on establishing and understanding of the details of this existing baseline.

KSC is a very data-rich environment. A goal-driven (as opposed to a data-driven) approach was selected in order to provide a structured methodology that was both effective and efficient. The methodology utilizes the accepted approach of functional decomposition. Functional decomposition takes a process-oriented approach, and breaks the KSC organization into major lines of business, referred to as functions. Based on center executive feedback, the following five broad functions capture the work that KSC employees are currently performing and will perform in the future:

1. Program management/operations/insight
2. Applied technology development
3. Enabling functions
4. Institutional/environmental management
5. Marketing and external activities

These top-tier functions were progressively broken down into subordinate functions, referred to as "processes" in the BPR literature. Each function and process was then mapped to a series of supporting matrices which linked such data elements as organizational responsibility, skills, customers, and roles and missions.

This was found to be analogous to the approach conducted in previous limited analyses at NASA Headquarters and at Stennis Space Center. In order to maintain agency consistency, many of the same terms and definitions utilized in these efforts were incorporated into the WAVE initiative. For example, the subordinate functions are referred to as "activities" in this research, instead of "processes" (as in the literature). The same glossary of terms used in these previous initiatives was incorporated into this project. It is to be emphasized, however, that these previous efforts focused more on a ZBR of Headquarters and Stennis, rather than a true BPR analysis. Since the objectives of this project were fundamentally different, the details of the approach diverged significantly from these earlier analyses.

As noted in Figure 6.2, a core team was established to manage and guide the overall processes associated with conducting this review. The critical task of the core team was to identify what occupational activities were being performed at that time, and to document (via input from KSC employees) what skills were currently available and those that will be required in the future. Additionally, the core team formulated recommendations to address any skills requirement gaps. These recommendations included retraining and redeployment of personnel. Teams were also established for each of the five functions identified. These functional teams further defined specific work activities associated with each

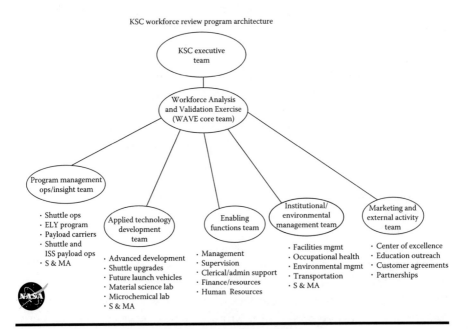

Figure 6.2 KSC workforce review program architecture.

of the functions. Each KSC directorate/office was asked to conduct data collection, using a survey tool distributed with instructions. This data collection effort encompassed a 30 day lead time. The functional teams then collected and synthesized the data for the core team. Both teams worked together to evaluate, clarify, and integrate the data. There was opportunity for feedback and the sharing of results with the organizations.

The core team then forwarded their recommendations to KSC senior executive management. Securing the support of senior executive management was critical for the success of the project, as well as to provide input for further definition of project scope, schedule, and deliverables. It was also critical to obtain their guidance regarding goals and priorities for the project.

As frequently identified in the literature (e.g., Knight and Wilmott, 2000; McNulty, 2004), there is always resistance to the changes that a reengineering project will introduce. In recognition of this, the organization and preparation stage was split into two phases. Phase A provided the foundation for characterizing the existing baseline and a first iteration of the skills gap. Phase B built on these first steps, and organized the true reengineering effort. Within the skills assessment phases, there may be some employee apprehension. Stakeholders in the status quo, and their interests, were then identified. Plans and methods were developed to obtain their "buy-in." Buy-in became more critical to the success of the later reengineering project phases. It was necessary to establish that the proper methods of communicating the intent and progress of the project are very important. Both NASA Glenn Research Center and Goddard Space Flight Center have undertaken similar initiatives. Analogous webpages were constructed to communicate project status and intent with KSC personnel.

Data management became increasingly important as the data collection proceeded. Matrices were constructed using Microsoft (MS) Excel spreadsheets. These spreadsheets were later translated into MS Access, which became the final repository for the data. MS Access is also the software used for the skills assessment software originally developed at Glenn. Selection of a common software platform was an important consideration for later reengineering phases.

6.3.2 Identification of the Existing Baseline

The purpose of the identification stage is to develop and understand a customer-oriented process model (Hammer and Champy, 2006). The identification stage provides definitions of customers, activities, as well as measures of performance and success, identification of value-adding activities, responsible organizations, resources available, and activity volumes, durations and frequencies. In a formal BPR study, the data collection requirements are quite extensive. For example, Maganelli and Klein (1994) provide more than 20 pages of representative types of data collection matrices. Conversation with KSC personnel indicated a sense of being "surveyed-out," after numerous types of surveys over the past few years.

- To KSC management:
  – Activities to function matrix.
  – Supporting matrices.
  – Skills inventory extension.
  – Survey: vision of future.
- To KSC employees:
  – Skills inventory.

Figure 6.3　Data collection approach.

Efforts were made to utilize existing data to the extent possible, and to limit the distribution of new surveys. The data collection strategy is depicted in Figure 6.3.

This first step in the data collection process was the identification of supporting activities for each of the five primary functions. The ISO 9000 processes, documented at KSC, served as the initial activity baseline. The individual blocks within the process flows indicated steps subordinate to the activity. The designated ISO process owners were contacted to confirm or update the flows. They were also encouraged to detail any processes/activities that were not already documented.

A listing of the confirmed activities was then distributed to the individual KSC directorates and offices, with a request that they identify the functions associated with each activity worked in that area. Data collection matrices were distributed as MS Excel e-mail attachments. This allowed for quick delivery and analysis. Each function–activity–step combination was assigned a unique identifier that was later used as a key field in the Access database, as well as the subordinate matrices.

Based upon the feedback regarding the Activity to Function Matrix, a set of supporting matrices were developed. Only two of these matrices were considered by the core team to be required: Roles and Missions, and Time. The Roles and Missions Matrix linked the following data:

1. Rows along Y-axis: Function, Activity, Step
2. Columns along X-axis: Responsible Organization, KSC Goal/Objective, External Customer, Internal Customer, External Supplier, Internal Supplier, Performance Criteria, Executive Priority

Much of this data already existed, and was initially filled in by the functional team for the organizations to confirm or modify. The ISO processes indicated an initial view of the responsible organizations. This data collection instrument largely confirmed that association. The type of involvement was also identified (i.e., "responsible for," "provides information," or "receives notification"), which was used in the subsequent phases of the BPR analysis. The organizational responsibility was identified at the appropriate level, either the activity or the step within the activity. The KSC Roadmap was used to select the relevant goal/objective for the activity/step. Performance indicators had already been mapped to the goals and objectives. The issue was to determine which of these indicators were to be

Linkage to be established:

Function
|
Activity ————————
| |
Step |
| |
Organization |
| |
Employee |
| |
Skill ————————

Figure 6.4 Linkage established.

considered critical success factors for KSC. The executive priorities were established earlier.

The Time Matrix identified the duration, volume of work, and frequency for each activity. This was alternatively reported at the step level, when there was sufficient data available. The matrix also assessed the impact of each activity/step on the performance measures of the process, in order to identify those activities/steps that added value (i.e., contributed to meeting a customer's want or need), those that did not, and those that were primarily for internal control. This matrix could be used to derive dollar values, if it was deemed necessary, by combining it with the other matrices of this effort in conjunction with the KSC Chief Financial Office databases. (This aspect may be investigated at a later date.) The purpose of the Time Matrix was to support the evaluation of alternatives, later in the BPR analysis, primarily through the use of process modeling software.

As indicated in Figure 6.4, the purpose of the data collection was to establish a linkage from the functions (indicating existing KSC primary business lines) to the individual skills of each employee. The existing KSC organizational charts were then consulted to provide the next link in this analytical chain (i.e., the employees for each organization). The organizational charts and associated Human Resources data also provided headcount and full time equivalent (FTE) values.

A skills assessment was then completed by the individual KSC employees. The Glenn Research Center software and approach were reviewed and recommended (with some minor modifications) for incorporation into this data collection. It is important to highlight that this skills assessment was intended not only to identify existing job-related skills, but also those skills that will be needed in the future.

6.3.2.1 Rough-Cut Skills Gap Determination

KSC senior executive management requested a rough-cut skills gap determination as an intermediate deliverable. The purpose of this task was to provide an initial approximation of the delta between existing skills and those needed in a specific future fiscal year for personnel retraining and redeployment. This analysis utilized the KSC employees' inputs to the Glenn skills inventory software as the existing skills baseline. A further modification of the software was required to permit management to identify their perceived future skill requirements. The quantity needed (in terms of FTE) per skill was entered. This was stored in an independent file from the existing skills baseline, and provided the objective skills baseline.

The gap analysis compared the existing and objective skills baselines. This was accomplished via an Excel spreadsheet interfaced to the two Access database files. The spreadsheet counted the number of existing employees with a specific skill set, and compared them to the total number required in the target fiscal year. A report indicating the resulting deltas by skill set was generated. It is critical to note that the objective skills file developed for this rough-cut analysis was insufficient for the subsequent BPR analysis steps. Further efforts, identified in subsequent sections of this chapter were necessary.

6.3.3 Organization and Preparation (Phase B)

Organization and preparation (Phase B) reoriented the project toward the next steps in the BPR effort. A process modeling software package was investigated and obtained at this time. Process modeling is a proven analytical tool, used to describe a business process by means of a workflow diagram, supporting text, and associated tools. Process modeling tools are classified by Hunt (1996) as

1. Flow diagramming tools: help define processes by linking text descriptions of the processes to workflow symbols.
2. Computer-aided software engineering (CASE): provide a conceptual framework for modeling hierarchies and process definitions. Usually built on a relational database, they provide functions that include linear, static, and deterministic analysis. These are considered the higher-end BPR softwares.
3. Simulation tools: provide capabilities to determine how the work flows through the process under specifically defined circumstances.

Hunt (1996) emphasizes that these tools do not necessarily provide the solution to BPR, but rather are technical aids that can assist in selecting the correct process change to make, and can help facilitate that change. She describes a variety of these software tools, and provides detailed guidance on choosing the process modeling tool appropriate to one's specific processes and budget. A customized software tool, developed at KSC, was selected for utilization on this project.

6.3.4 Determination of Specific Areas for Transformation

As originally discussed by Hammer and Champy (1993) and reinforced by subsequent BPR books and articles by other authors, the next step is to clearly identify the specific areas for further analysis and reengineering. An organization cannot restructure everything at once. As indicated by numerous case studies in the literature (e.g., Knight and Wilmott, 2000), to do so is a prescription for project failure. The methodology for determining specific areas for transformation required the following data to be obtained from the individual directorates and project offices:

1. Identify set of new customers and their requirements.
2. Add new activities or modify existing activities in order to support new customers.
3. Identify set of existing customers whose method of support will change.

For these situations, a linkage to other activities was established through the review of the Roles and Missions Matrix. The internal/external supplier of one activity is either a previous activity or the internal/external customer identified for that activity. A network of such activities was developed. A change in customer, or in the method of customer support, may cause changes in these subordinate activities, which were then documented. It was also important to highlight any activities in the original linkage that would no longer support the revised customer base upon transition to research and development (R&D).

6.3.4.1 Core Competencies

The critical subset of these activity chains dealt with the center's core competencies and core capabilities. Briefly stated, they are those activities that are considered most vital to the organization, since they are both strategic and value-added. At KSC, they are the activity chains that directly affect the Roadmap's stated goals and objectives (Bridges, 1998). It is this subset of the entire list of activities that provides the focus for subsequent reengineering.

The subject of core competencies has received considerable attention in the management literature (e.g., Simpson, 1994; Cooper, 2000; Drejer, 2002). As defined by Gallon et al. (1995), "core competencies are those things that some companies know how to do uniquely well." Core competency statements are normally at a very summary level. They exist at the top of an organizational hierarchy. Organizations encompass a variety of discrete activities, skills, and disciplines that Gallon et al. (1995) refer to as primary capabilities. These primary capabilities are the responsibility of individual departments within the organization.

A subset of these primary capabilities has a direct and significant impact on the organization's success. This subset is referred to as core (or critical) capabilities. In commercial business organizations, these core capabilities are responsible for improved products or service, reduced costs, improved speed to market, and competitive advantage. Core competencies may be considered to be aggregates of these core capabilities. For example, core capabilities in surface coating formulation and continuous coating processes provide the basis for 3M corporation's core competency (Simpson, 1994).

KSC's core competencies in launch and payload processing were clearly defined in the KSC Implementation Plan and accompanying Roadmap (Bridges, 1998). A variety of parallel initiatives were underway in order to identify the underlying core capabilities and primary capabilities, with special emphasis on identifying the skills associated with them. This project assignment specifically dealt with the development of a KSC competency assessment and skills inventory system.

According to Cooper (2000), a skills inventory "in its simplest form is a list of the names, certain characteristics, and skills of the people working for an organization. It provides a way to acquire these data and makes them available where needed in an efficient manner." Good skills inventories enable organizations to quickly determine the types of people and skills that are available for future projects.

The establishment of the skills inventory database was merely the first step in developing the manpower planning software that KSC requires. Although the prototype database could generate a variety of output using MS Access's report generation capabilities, an intelligent user interface would need to be created and front-ended to the database to more efficiently and effectively identify candidate employees with the requisite skills. The Resource Management System (RMS) is a web-based software system developed by the Advanced Information Systems (AIS) team at KSC. It is designed to provide project managers with a tool to more effectively manage their resources, specifically people and budgets. The AIS effort to develop an RMS with a skills tracking/training component complemented the center's desire to develop a core competency system. Allowing the skills inventory database to reside within RMS created a common platform for KSC use.

6.3.5 Transformation of Target Areas into the Objective

The purpose of this stage was to specify the technical and human dimensions of the new process. The activities targeted for investigation were reviewed in detail, and alternative recommendations were generated and evaluated. According to the literature, the initial set of targets should be near-term opportunities, those that can be resolved over the next 6 month period (Simpson, 1994). These targets should be core capability activities currently in use, and that will also be used upon transition to develop. KSC activities were investigated in sequence by the associated executive priority assigned, as well as the perceived probability for reengineering success. The idea was to demonstrate the benefits of the project as early as possible (Drejer, 2002).

Longer-term activities were addressed in subsequent 6 month blocks. These primarily focused on the remaining core capability activities. These longer-term activities were investigated with consideration of their links to other activities (i.e., the activity chains or activity clusters). Reengineering an entire activity cluster in one time block avoided creating "islands of efficiency" among process bottlenecks, and better leveraged project efforts. Again, these were sequenced consistent with the associated executive priority. At this point, any activities that were identified as not supporting the R&D environment were discarded.

6.3.5.1 Analysis of Scheduled Activities

The activity linkages often offer areas for improvement. Consideration was given to the potential that the re-sequencing of steps in the activities, or activities in the chain, or the reassignment of organizational responsibilities would improve performance. Similarly, consideration was also given to the identification of instances where better coordination among activities within a cluster would improve performance. This analysis also defined changes needed to reduce or simplify organizational interfaces, both internal and external. Duplication of information flow was identified and eliminated, as were other types of redundancies.

The analysis also identified alternatives that reduced the number of non-value-added activities or steps (as indicated by the Time Matrix) by simplifying the control structure of the activity. This was done by integrating the controls into the value-added activity/step by replacing error detection with error avoidance techniques. Actual error detection was moved closer to the point of the occurrence of the error. This task also looked at the logical relationship among the activities in order to find opportunities to perform activities/steps in parallel that were conducted in series. This improved the speed of the overall activity or activity cluster. The application of new technologies to the activities was investigated at this time.

6.3.5.2 Evaluation of Alternatives

The alternatives generated during the previous analysis step were then evaluated based on their resulting change in the performance indicator identified in the Roles and Missions Matrix for this activity. Improvements in time/cost efficiency were projected by utilizing the process modeling software. The original activity flows, time matrix values, and organizational resources (in terms of headcount or FTE) were input as the basis of the existing model. This existing model was executed, and the results of the simulation reviewed. The existing model was then modified in accordance to the alternative being evaluated. This revised activity model was subsequently simulated in order to determine any improvements in time, manpower, and budget requirements. The alternative, and its associated requirements delta to the original, was later evaluated by the functional and core teams. Evaluation of large-scale changes (beyond a pre-defined threshold) was forwarded to a "blue-ribbon" panel designated by senior executive management. The alternative would then be accepted, rejected, or recommended for revision.

6.3.5.3 Comparison of Accepted Alternatives to Original Baseline

For each alternative accepted, the resulting impact on the KSC workforce was determined. This was accomplished by comparing the revised activity with the original. Based upon the changes made to the activity and its subordinate steps, the associated Roles and Missions Matrix and Time Matrix were formally updated. The associated delta for the specific skills required and headcount could now be defined. The results of this process provided a much more accurate forecast of the future skills gap that the rough-cut skills gap analysis described earlier.

6.3.5.4 Organizational Change

The results of this workforce analysis provided the opportunity for organizational change. Organizations normally grow with limited consideration of the core activities/processes that they were originally established to support. Inefficiencies frequently result. At this point in the BPR process, the necessary insights had been developed to guide a restructuring along activity/process lines. The overall management structure is determined by integrating the objective baseline with input from KSC personnel as well as benchmarking other organizations.

At the time of this project, KSC was organized much like a pure product organization in commercial industry (Kerzner, 2006). For example, Space Shuttle processing constituted one organization, while Expendable Launch Vehicles was another. One alternative structure that was suggested was a matrix structure, where the functional departments provide resources to the individual projects. The current KSC organizational structure (NASA, 2006) has begun to adapt to this suggested approach. The matrix organizational structure provides the flexibility for further organizational opportunities (Kerzner, 2006). For example, if the Glenn software is used agency-wide, it could provide the basis for a 3D matrix. Here, the individual functional departments in multiple centers (e.g., Johnson, Kennedy, and Goddard design engineering departments) could provide people to support the Constellation program efforts in a virtual organization across geographic boundaries.

Whichever management structure is recommended is then used as the summary framework for establishing the organizational boundaries. This task considers changing the organizational structure in order to reduce the number of boundaries traversed by a process. Thus, the organization is established around specific activities. This will result in consolidation of resources and streamlined processes.

Once the objective organizational structure is established, consideration must then be given to the issues of retraining and redeployment. The associated skill needs probably will not match the existing personnel in the modified organization. The skill background of each employee in the organization would be reviewed for the potential for retraining. Employees associated with activities no longer required in the objective environment then provide a candidate pool for redeployment and/ or retraining. Retraining planning, for employees either within or outside the organization, can be based on a weighted change in acquisition of skill level. This technique assigns weights to each existing and future skill and associated level of expertise (e.g., Maganelli and Klein, 1994). The weighted values are then summed for each candidate being considered. This produces a measure of the difficulty in making the transition from an employee's existing skill base to the required one. The smaller the transition difficulty measure, the easier the retraining and the better the candidate for this specific slot.

6.3.5.5 Implementation

The first step toward the actual creation and implementation of the objective was to determine the methods to be used for system validation. These may include parallel or pilot testing. The revised activity may be tested in parallel to the existing method in order to confirm that it works as expected. Parallel testing normally requires that redundant resources be committed to the activity. In a resource-constrained environment like KSC, this was not possible. Establishing a limited pilot line was a more suitable approach to evaluate the revised activity prior to full-scale implementation.

At the same time that validation is occurring, the methods for conversion and transition need to be developed. The revised activity may require modification of supporting information systems, databases, hardcopy documentation, and organizational procedures. These support mechanisms for the original activity may present roadblocks to the effective implementation of the new activity. An operations-ready version of the new activity must include a time-phased deployment plan, including the development and testing of supporting databases, the development and testing of systems and procedures, documentation, and data conversion, as well as the actual implementation of the revised activity.

Based upon the detailed deployment plan, the new activity is operated in a limited area to support validation. Any flaws are identified and corrected in a controlled manner until full implementation occurs. Subsequent to implementation, the activity is periodically monitored, reanalyzed, and fine-tuned, so that the further performance gains may be realized.

6.4 Conclusions

The Consolidated Appropriations Act of 2008 (PL110-161) directs that "the NASA Administrator shall prepare a strategy for minimizing job losses when the National Aeronautics and Space Administration transitions from the Space Shuttle to a successor human-rated space transport vehicle" (NASA, 2008b). This transition is the most challenging and complex for the agency since the end of the Apollo program and the beginning of the Space Shuttle program. As noted by Gerstenmaier and Gilbrech (2008), NASA formally defines its approach to transition as "the careful planning, optimized utilization, and responsive disposition of processes, personnel, resources, and real and personal property, focused upon leveraging existing Shuttle and ISS assets for Exploration program's safety and mission success." A number of interrelated initiatives have been conducted or are currently underway at KSC, designed to support the center's transition from Space Shuttle operations to the Constellation development program. BPR provides a structured methodology to apply systems engineering concepts to the complexities of organizational workforce planning. This chapter addresses the development of a BPR plan for the restructuring, retraining, and redeployment of elements of the KSC workforce during the transition. This plan has been partially implemented, and may provide a template for future use of BPR in agency workforce planning.

Author

Gary P. Moynihan is a professor in the Department of Civil, Construction, and Environmental Engineering, the University of Alabama. He received his BS (chemistry) and MBA (operations management) degrees from Rensselaer Polytechnic Institute, and his PhD (industrial engineering) from the University of Central Florida. While at the University of Alabama, he has taught, developed, and expanded courses in information systems development, project management, and engineering economics. His primary areas of research specialization are information systems design and analysis, project management, and operations analysis. Dr. Moynihan has served as principal or co-principal investigator on research grants funded by NASA's Johnson Space Center, Marshall Space Flight Center, and Kennedy Space Center, the U.S. Army Aviation and Missile Command, the Federal Aviation Administration, the U.S. Department of Energy, and Mercedes-Benz. Prior to joining the University of Alabama, Dr. Moynihan was employed for 10 years in the aerospace industry, where he was a manufacturing engineer, industrial engineer, and a systems analyst for a variety of company programs, including missile defense and fixed wing/rotorcraft avionics. He has also held production supervisor positions in the computer and chemical processing industries.

References

Bambarger, B. 1994. Corning Asahi Video Products Co. eliminates cost of errors for $2 million savings. *Industrial Engineering* 25: 28–30.
Blanchard, B. and Fabrycky, W. 1998. *Systems Engineering and Analysis* (3rd edn.). Upper Saddle River, NJ: Prentice Hall.
Bridges, R.D. 1998. Kennedy Space Center implementing NASA's strategies. Kennedy Space Center, FL.
Brittain, C. 1994. Reengineering complements BellSouth's major business strategies. *Industrial Engineering* 25: 34–36.
Cabana, R.D. 2009. Message from the desk of the center director. http://www.nasa.gov/centers/kennedy/about/cd_welcome.html

Cooper, K. 2000. *Effective Competency Modeling and Reporting*. New York: American Management Association.

Drejer, A. 2002. *Strategic Management and Core Competencies*. Westport, CT: Greenwood Publishing Group.

Gallon, M.R., Stillman, H.M., and Coates, D. 1995. Putting core competency thinking into practice. *Research Technology Management* 38: 20–28.

Gerstenmaier, W. and Gilbrech, R. 2008. Workforce transition strategy initial report: Space shuttle and constellation workforce focus. Kennedy Space Center, Orsino, FL.

Hammer, M. and Champy, J. 1993. *Reengineering the Corporation*. New York: HarperBusiness.

Hammer, M. and Champy, J. 2006. *Reengineering the Corporation: A Manifesto for Business Revolution*. New York: Harper-Collins.

Hunt, V.D. 1996. *Process Mapping*. New York: John Wiley & Sons.

Kerzner, H. 2006. *Project Management: A Systems Approach to Planning, Scheduling and Controlling* (9th edn.). New York: Wiley.

Khosrowpour, M. 2006. *Cases on Information Technology and Business Process Reengineering*. Hershey, PA: Idea Group Inc.

Knight, D. and Wilmott, H. 2000. *The Reengineering Revolution: Critical Studies of Corporate Change*. Thousand Oaks, CA: Sage Publications.

Maganelli, R.L. and Klein, M.M. 1994. *The Reengineering Handbook*. New York: American Management Association.

Malhotra, Y. 1998. Business process redesign: An overview. *IEEE Engineering Management* 26: 27–31.

McCloud, J. 1993. McDonnell Douglas saves over $1,000,000 per plane with reengineering effort. *Industrial Engineering* 23: 27–30.

McNulty, T. 2004. Redesigning public services: Challenges of practice for policy. *British Journal of Management* 14: S31–S45.

National Aeronautics and Space Administration (NASA). 1997a. NASA workforce restructuring plan. http://www.hq.nasa.gov/office/codef/codefm/rs97plan.html

National Aeronautics and Space Administration (NASA) Office of Human Resources and Education. 1997b. Human resources systems replacement study (draft). Kennedy Space Center, FL.

National Aeronautics and Space Administration (NASA). 2004. The vision for space exploration. Kennedy Space Center, FL.

National Aeronautics and Space Administration (NASA). 2006. Kennedy Space Center organizational chart. Kennedy Space Center, FL.

National Aeronautics and Space Administration (NASA). 2008a. Kennedy Space Center's annual report FY2007. Kennedy Space Center, Florida.

National Aeronautics and Space Administration (NASA). 2008b. Workforce transition strategy initial report. Kennedy Space Center, FL.

Osborne, D. and Hutchinson, P. 2004. *The Price of Government: Getting the Results We Need in an Age of Permanent Fiscal Crisis*. New York: Basic Books.

Simpson, D. 1994. How to identify and enhance core competencies. *Planning Review* 22: 24–26.

Skytte, K. 1994. Engineering a small system. *IEEE Spectrum* 31: 63–65.

United States General Accounting Office Accounting and Information Management Division. 1997. *Business Process Reengineering Assessment Guide (Version 3)*. Washington, DC: General Accounting Office.

White, J. 2007. *Managing Information in the Public Sector*. Armonk, NY: M.E. Sharpe.

Chapter 7

Systems Engineering Case Studies

Charles M. Garland and John Colombi

Contents

Case studies provide unique and necessary insights, as well as a realistic practical component, of systems engineering (SE) knowledge. In this chapter, the authors present the stories of early design and development of several large complex systems, spanning three decades from the Department of Defense (DoD) and the National Aeronautics and Space Administration (NASA). These systems include the F-111 sweep-wing fighter, the large front-loading C-5 mobility aircraft, the serviceable Hubble Space Telescope (HST) satellite, the cold war Peacekeeper missile system, the A-10 "tank-buster" aircraft, the stealth B-2 bomber, and the global positioning system (GPS) constellation. Each system provides the backdrop for discussion of several SE learning principles, originally depicted in a framework by George Friedman and Andrew Sage: Experience the technical decision making, requirements definition, logistics support planning, verification, risk mitigation, configuration management, and many other SE activities within the acquisition environment and political context of these historic systems.

7.1 Case for the Case Study

Case studies have long been used as a tool in academia, to support a more student-participative pedagogy. The pragmatic quote by 1869 Harvard President Charles Elliot is used to this day, "Our major problem is not what to teach, but how to teach it." This issue concerns not just "how to teach" but perhaps the more important aspect of "how to learn." A wide range of professional schools, including Harvard's law, business, and medical schools, have concluded that the best way to teach realistic life skills is by the case method (Garvin, 2003). So, case studies and the case study method offer a potentially better way to address "how to learn." An argument can be made that such methods can be especially useful to teach both the practice and art of systems engineering (SE) and technical decision making.

The SE process, used in today's complex projects, has matured based on well-founded principles and heuristics from past developments. Application of SE across real programs, both past and present, can provide a wealth of lessons to support the teaching of the disciplines' foundational principles. Cases can facilitate learning by emphasizing the long-term consequences of the SE and programmatic decisions on program success. SE case studies can assist in discussion of both successful and unsuccessful methodologies, processes, principles, tools, and decision material to assess the outcome of alternatives at the program/system level. In addition, the importance of using skills from multiple professions and engineering disciplines and collecting, assessing, and integrating varied functional data can be emphasized. When taken together, the student reading case studies can be provided real-world, detailed examples of how the process attempts to balance cost, schedule, and performance. Case studies should be used to illustrate both the utilization and misutilization of SE learning principles and allow the reader to infer whether

1. A balanced and optimized product was provided to a customer or user
2. Effective requirements analysis was applied
3. Consistent and rigorous application of SE management was applied
4. Effective test planning was accomplished
5. Major technical program reviews were rigorous
6. Continuous risk assessments and management were implemented
7. Cost estimates and policies were reliable
8. Disciplined application of configuration management was used
9. A well-defined system boundary was defined
10. Problem solving incorporated understanding of the system within larger environmental context (customer's customer)

SE case studies can also highlight other aspects of a system development. One of these aspects could be the communication between various stakeholders. For large government and defense systems acquisition, these important communications comprise the following:

1. The interface between the program manager and the systems engineer, which is essential to balancing cost, schedule, and technical performance of the program.
2. The interface between the government (acquisition office) and the contractor to translate and allocate performance requirements into detailed system requirements and design.

3. The interface between the acquisition office and the user (or customer) to translate operational needs into requirements.

Lastly, the audience for a case must be considered. A case may be read predominantly by systems engineers and practitioners who want the executive summary and explicit learning principles. A case may be used in undergraduate or graduate education, with learning principles implicitly discovered by the participants in the classroom. Thus, a framework may be necessary to guide the format to best present aspects of SE or SE management.

7.2 Framework for Case Analysis

George Friedman and Andy Sage (2004) offer such a framework to guide SE case writing. The Friedman–Sage (F–S) framework presents a nine-row-by-three-column matrix as shown in Table 7.1. Six of the nine concept domain areas represent phases in the SE life cycle:

1. Requirements definition and management
2. Systems architecting and conceptual design
3. Detailed system and subsystem design and implementation
4. Systems and interface integration
5. Validation and verification
6. System deployment and post-deployment

The remaining three areas represent necessary process and systems management support:

7. Life cycle support
8. Risk assessment and management
9. System and program management

While other concepts in SE could have been identified, this framework suggests that these nine are the most relevant to SE in real practice; they cover the essential life cycle processes in systems acquisition and the systems management support. Most other concept areas that could be identified would probably be a subset or closely related to one of these. The three columns of this two-dimensional framework represent the responsibilities and perspectives of the government and the contractor and the shared responsibilities between the government and the contractor. In their paper, Friedman and Sage (2004) highlighted various concepts within each of the 27 elements of the framework as requirements for good SE. The following summarizes salient points of those elements by concept domain and responsibility domain.

Table 7.1 Friedman–Sage (F–S) Framework

Concept Domain	Responsibility Domain		
	Government	*Contractor*	*Shared*
Requirements definition and management	Government shall integrate the needs of its user organizations with the management activities of its development organizations	Requirements shall flow down in a coherent and traceable manner from the top level to all lower levels	Customer and contractor shall share with one another their knowledge of the state of technical maturity
System architecture and conceptual design	Systems architecture shall be employed by the government	Systems baseline architecture of complex programs shall be established early	Systems architecture should be established early
System and subsystem detailed design and implementation	Customer shall share high-level measures of effectiveness with the contractor	System design shall proceed in a logical and orderly manner	Government customers and contractors shall have a contractually feasible sharing of system design responsibility
Systems and interface integration	Government shall assure that all its operational systems are compatible and mutually supportive in a broad "system of systems" context	Contractor shall assure that systems integration supports total systems functionality	Contractor and government shall assure that all systems are interfaced with other existing operational equipment
Validation and verification	Government shall be the final word on the confidence levels derived from its testing	Every requirement shall have a test and every test shall have a requirement, which requires validation and verification	Test criteria shall be shared early

Deployment and post-deployment	Government shall assure that a properly funded operational evaluation occurs	Contractor shall maintain the appropriate engineering and testing organizational capabilities to support recommending possible changes in the system design or support	Government and contractor shall work cooperatively to conduct effective operational test evaluation
Life cycle support	Funding support, throughout a program's life cycle, shall be maintained in balance	All design activities shall be performed from the viewpoint of the entire life cycle	Balanced blend of all methods, measurements, technologies, and processes shall be employed in support of an effective life cycle
Risk assessment/ management	Risk management at all program levels shall be an essential and inherent part of all life cycle activities	At every level of detail, risk shall be identified, prioritized, and mitigated	It is to the government's benefit as well as the contractor's to identify and mitigate risk early
System and program management	Government shall establish security levels to protect special operational capabilities	Each and every systems engineering program shall have a systems engineering management plan	The role of systems engineering in program development and management shall be recognized and supported

Source: Modified from Friedman, G. and Sage, A., *J. Syst. Eng.,* 7, 84, 2004.

7.3 Systems Engineering Vignettes

Primary source material for the following case vignettes was taken from a series of SE case studies published by the Air Force Center for Systems Engineering (USAF, 2009). The authors of this chapter managed and edited several of the final cases. Each case conveys the story of the program and delineates a set of several learning principles. One relevant example for each of the nine concept domains of the F–S framework, along with supporting details, is conveyed in the following vignettes. A set of questions to the reader has been included in each vignette to encourage thoughtful reflection.

These Department of Denfense (DoD) and NASA programs span over three decades. While the exact SE practices, life cycles milestones, and government policies have varied over the years, they remarkably share much commonality in general process. Concepts get developed, refined, and matured. Technology and prototyping are often examined early in support of the proposed system. Requirements get written, rewritten, and debated. A solicitation goes out from a program office, with defense and national contractors or teams bidding on the work; the size of these projects must not be forgotten. They range from hundreds of million (U.S. dollars) to billions. Once a contractor or team is selected, design is matured through preliminary and critical design reviews toward development and/or full-scale production. For defense weapon systems, the DoD Instruction DoD I 5000.02 (updated almost 20 times over three decades) currently guides overall acquisition procedures, reviews, products and plans, and oversight.

7.3.1 Requirements Definition and Management Concept Domain

In a sound SE approach to a large, complex system development, the system requirements and specifications should be thoroughly evaluated for effectiveness, cost, and technical risk, before beginning the program. Thereafter, they should be managed throughout the development program. While reviewing lessons from a variety of DoD acquisitions, requirements and requirements management unequivocally appears as a problem or a major challenge.

7.3.1.1 F-111—In the Beginning

Three streams of activity collided during the earliest days of the administration of the 35th President of the United States, John F. Kennedy. First, President Kennedy chose Robert S. McNamara to become the 8th Secretary of Defense. McNamara was one of ten men who served together during World War II (WWII) in a statistical control group of the Army Air Forces. After the war, the 10, who later became known as the "Whiz Kids," were hired by Henry Ford II. They brought a statistical-based management style to Ford. McNamara's goal as Secretary of Defense was "to

Figure 7.1   F-111F. (Photo courtesy of USAF.)

bring efficiency to a $40 billion enterprise beset by jealousies and political pressures while maintaining American military superiority" (John F. Kennedy Presidential Library & Museum, 2009). Second, in 1958, the U.S. Air Force (USAF) established operational requirements for a new fighter-bomber that was to surpass the capabilities of aircraft in the inventory such as the F-100 Super Sabre and the F-105 Thunderchief (Richey, 2005). These requirements were refined over the next 15 months culminating in specific operational requirement #183 that called for an aircraft that could achieve a speed of Mach 2.5 at high altitude and possess the capability to travel 400 miles at Mach 1.2 at low level. Third, in December of 1960, the U.S. Navy (USN) F6D Missileer program was cancelled in part because it was viewed as too costly. The subsonic F6D aircraft, armed with long-range high-speed air-to-air missiles, was slated to replace the F-4 Phantom and the F-8 Crusader in the role of naval fleet air defense (FAD) fighter.

The collision of these three activities resulted in the beginning of a development program to provide a common aircraft solution to meet multiservice requirements. The resultant aircraft would become the F-111 (Figure 7.1).

7.3.1.2 F-111 Aardvark Program

Upon entering office on January 21, 1961, it did not take Secretary McNamara long to make progress toward his efficiency goals. In February of 1961, he directed that the services study the development of a single aircraft that would satisfy the mission requirements of both the Air Force and the Navy. He also directed that the program meet the Army and Marine Corps needs for a close air support (CAS) aircraft. Thus, the tactical fighter experimental (TFX) project was born. Making things even tougher, the USAF and USN variants were required to be 80% common in terms of structural weight and parts count. The services at an early stage convinced Secretary McNamara that the CAS mission requirement could not be satisfied by the TFX, and the Marine Corps and the Army were dropped from the program. Despite serious service doubts that commonality could be achieved, he adamantly directed the Air Force and Navy to combine their requirements in a single request for proposal (RFP).

7.3.1.3 Questions to the Reader

What do you believe could make commonality very difficult to achieve?
What could be some technical benefits resulting from the requirement to be 80% common?
What could be some nontechnical benefits of such commonality?

7.3.1.4 Disagreement from the Start

Combining requirements proved a difficult challenge for the Air Force and Navy. The Navy favored side-by-side seating for its FAD fighter, whereas the Air Force preferred tandem seating. The Navy wanted an aircraft equipped with a long-range search and intercept radar having a dish 48 in. in diameter, whereas the Air Force needed an aircraft equipped with a terrain-following radar optimized for low-altitude operations. The Navy needed an aircraft that was optimized for long loiter times at medium to high altitudes at subsonic speeds, whereas the Air Force insisted on an aircraft capable of low-altitude operations and supersonic dash performance. Undaunted, Secretary McNamara pressed forward with the project and directed that the Air Force would be the lead service for the development of a common TFX aircraft.

By August of 1961, the Secretary of the Navy reported to Secretary McNamara that the compromise TFX design could not meet the Navy requirements. The Air Force desired an aircraft weighing less than £75,000 gross, while the Navy needed the gross weight to be kept below £50,000. In addition, carrier operational requirements necessitated that the overall length be kept below 56 ft so that it could fit aboard existing carrier elevators. McNamara ordered the Navy to accept a design sized to accommodate a 36 in. radar rather than the 48 in. radar and to accept a gross takeoff weight of £55,000.

On September 29, 1961, an RFP was issued to Boeing, General Dynamics, Lockheed, Northrop, Grumman, McDonnell, Douglas, North American, and Republic. The Air Force's version of the TFX was to be designated F-111A, with the Navy's version being designated F-111B. The difficulty by the contractors in proposing a solution was reflected in the fact that it took four separate proposal submittal rounds before the services made a choice for a winner. Air Force and Navy leadership preferred the Boeing design. In a controversial reversal of service preference, Secretary McNamara awarded the contract to General Dynamics citing their designs promised greater degree of commonality.

The F-111A and B aircraft shared the same primary structure, the same fuel system, the same pair of Pratt & Whitney TF30-P-1 turbofans, and the same two-seat cockpit in which the two crew members sat side-by-side. The side-by-side seating was a concession to Navy demands. The Navy also insisted that the cockpit be capable of doubling as an escape capsule for the crew in the case of an emergency, thus providing a parachute and the ability to float on water. The Navy F-111B's nose was 8 ft 6 in. shorter than the F-111A's because of the need of the aircraft to fit on existing carrier elevator decks and had 3 ft 6 in. extended wingtips in order to increase the wing area so that the on-station endurance time would be improved. The Navy version would carry a Hughes AN/AWG-9 pulse-Doppler radar and an armament of six Hughes Phoenix missiles, which had both evolved from the F6D program. The Air Force version would carry the General Electric AN/APQ-113 attack radar, the Texas Instruments AN/APQ-110 terrain-following radar and an armament of internal and external nuclear, and conventional air-to-ground stores.

7.3.1.5 Navy Steps Away from the F-111

The first Naval preliminary evaluation was held at Naval Air Testing Center Patuxent River in October of 1965. The F-111B was already in trouble since it was seriously overweight. Takeoff weight for a fully equipped aircraft was estimated at nearly £78,000, well over the upper limit of £55,000 as required by the Navy. The problems with the overweight F-111B were so severe that General Dynamics and Grumman were forced into a Super Weight Improvement Program (SWIP), most of the changes being incorporated into the fourth and subsequent F-111Bs. The fourth F-111B was fitted with an escape capsule in place of the individual ejector seats that were fitted to the first three F-111Bs. However, the fitting of this capsule more than offset the weight reductions achieved by the SWIP, and the F-111B remained grossly underpowered. Range was also below specifications and could only be increased by adding more fuel, making the aircraft even heavier. By October 1967, the Navy was convinced that the F-111B would never be developed into a useful carrier aircraft and recommended that the project be terminated. In May of 1968, both houses of Congress refused to fund F-111B production. McNamara had left office on February 29, 1968. A few months later in July, a stop-work order was issued to General Dynamics, after $377 million had been spent on the program.

7.3.1.6 Successful Aircraft

Despite the Navy's cancellation of their F-111B variant, the F-111 aircraft went on to prove a valuable asset for the Air Force. Six other variants were subsequently built. F-111D, E, F, FB-111A, and EF-111A variants were used by the United States. The Royal Australian Air Force bought an F-111C variant. The F-111 was retired from service in 1996 after compiling an incredible safety record of 77 aircraft lost in one million flying hours. In addition, this aircraft contributed successfully to Operation Desert Storm where F-111s flew 2500 sorties, dropped 5500 bombs, and destroyed 2203 targets.

7.3.1.7 Learning Principle

Ill-conceived sets of requirements (and specifications) should be identified early; else system development can become risky, difficult to manage, and difficult to achieve.

In the case of the F-111, the development of a set of specifications based on service requirements was seriously flawed. The fundamental issue was the disparate requirements for speed, altitude, range, and weight forced for the seaming benefit of commonality alone. The Air Force requirement was for a low-altitude penetrator (at Mach 1.2) and high-altitude supersonic (Mach 2.5) fighter/interceptor. The Navy requirement was for a subsonic fleet defense interceptor that could operate at long distances for extended periods of time to detect and destroy enemy aircraft outside the range at which they could launch antiship missiles.

7.3.2 Systems Architecting and Conceptual Design Concept Domain

The system concept and preliminary design must follow, not precede, mission analysis. Developing the system concept and architecture is one of the most important responsibilities of the systems engineer. It is a very creative process, and there is often no unique solution to satisfy the user requirements. This step helps guide and frame system development. The conceptual formulation phase of the A-10 military aircraft will be used throughout this section.

7.3.2.1 Close Air Support

CAS is defined as "air action by fixed or rotary winged aircraft against hostile targets that are in close proximity to friendly forces, and which requires detailed integration of each air mission with fire and movement of these forces" (Joint Chiefs of Staff, 2009). For CAS, fixed or rotary winged aircraft are in very close proximity to both friendly and enemy troops, and the integration of CAS firepower with the plans and movement of ground forces is extremely important. In the early 1960s, the Air Force believed that fighters that were not otherwise engaged could take on CAS when needed (Jacques and Strouble, 2008). The Air Force emphasized airplanes with speed for survivability and believed that a pilot trained to perform air-to-air combat could easily perform the air-to-ground mission. Additionally, it was believed that a dedicated CAS attack plane would be limited in capabilities and very vulnerable. Thus, the trend was for larger, faster multi-role fighter aircraft. The Army, on the other hand, wanted an aircraft that could carry a great amount of ordnance, loiter in the area for some time with excellent maneuverability, and had the ability to take hits from enemy ground fire. For years, this debate would continue, as to which service was responsible for the CAS mission and what type of aircraft would be best.

During the Vietnam War, the Air Forces primary fighter, the F-105, was big and fast, but was limited in its ability to fly closely and slowly enough to see the target, to work safely in poor weather, to carry sufficient ordnance, and to remain over the battle area. Eventually, the F-4 would be brought to the fight, but again, this aircraft did not have the low-speed, low-altitude, and loiter capability needed for CAS. A better interim solution was found with the use of the semiobsolete Navy A-1 Skyraider. Also, during the Vietnam War, the Army began employing armed helicopters to provide CAS to friendly ground forces. The AH-56A Cheyenne was designed as a large, fixed-wing, and rotary-wing aircraft with sophisticated avionics and a greatly increased capacity for attacking ground targets. The AH-56A was slow (maximum speed of 214 knots) and was armed with a 30 mm automatic gun in the belly turret and a 40 mm grenade launcher (or 7.62 mm Gatling gun) in the chin turret. It also carried antitank missiles and rocket launchers. The Army's transition into a role normally provided by the Air Force, namely, CAS, created tension between the services.

In 1965, a series of Air Force and Office of Secretary of Defense (OSD) studies examined CAS solutions. Concepts included modification of the A-7 aircraft as well as the F-5 fighter (Figure 7.2). Neither of these aircraft would be optimized for the close support role but could fill an interim capability gap. It was decided to pursue acquisition of the A-7D, a new variant of the Navy A-7 aircraft. The A-7D was expected to be low cost (about $1.5M/aircraft) and quickly obtainable, and the OSD authorized the Air Force to begin acquisition in December 1965. Subsequently, Navy and Air Force desires to improve the system design with new

Figure 7.2   Early competitors for the close air support mission. Army's AH-56 Cheyenne (left), the A-7D (top right), and F-5 (bottom right) (From Jacques, D.R. and Strouble, D.D., *A-10 Thunderbolt II (Warthog) Systems Engineering Case Study*, Air Force Center for Systems Engineering, Wright-Patterson AFB, OH, 2008, http://www.afit.edu/cse/ (accessed December 9, 2009.)

engines and avionics increased cost to about \$3.4M/aircraft, resulting in greatly reduced numbers of A-7Ds produced.

Additional CAS trade studies finally provided the way ahead for the Air Force. To fulfill the requirements for beyond 1970, the Air Force would take immediate steps to develop a new specialized CAS aircraft, simpler and cheaper than the A-7, and with equal or better characteristics than the older A-1. On September 8, 1966, the Air Force was authorized to design, develop, and produce a specialized CAS aircraft, designated the A-X (for Attack aircraft-eXperimental). This would eventually become the A-10.

7.3.2.2 Early Concept Studies and Mission Analysis

An early requirements package, called a requirements action directive, got the Air Force to contract for A-X concept definition. This requirements package cited an urgent need with an initial operational capability of 18 aircrafts delivered by 1970 and dictated maximum employment of existing state-of-the-art technology in its design. This would allow for a compressed conceptual design phase with lower risk. The A-X would be designed to provide CAS of ground units, escort of helicopters and low-performance aircraft, protection of landing surface forces and vehicle convoys, and armed reconnaissance.

The A-X was to be a single-man, lightweight aircraft with sufficient range and capacity to carry a maximum payload at low altitude from a main operating base to a forward area with a significant loiter time on station. Maneuverability requirements stressed agility in attack and reattack maneuvering at low speed, and the A-X required stability throughout a weapon release speed of 200–400 knots.

The A-X requirements directive called for fixed, internally mounted guns with a "capability equal to or better than four M-39 20mm guns." Survivability from ground fire was an essential characteristic for the A-X. Structural and system design would need to provide inherent survivability, to include self-sealing fuel tanks and, if power flight controls were used, a manual backup system would be provided. The pilot and critical flight systems would be protected from 14.5 mm projectiles (a common enemy antiaircraft shell). The aircraft was to also incorporate maintainability characteristics, allowing a minimum of maintenance effort and expenditure across its life cycle.

The A-X was to use an existing state-of-the-art engine in order to achieve an early initial operational capability (IOC). It would also use existing state-of-the-art equipment for avionics (communications, navigation, and weapons delivery systems). Communication equipment was to be compatible with forward air control equipment and Army airborne vehicles. Navigation equipment was to be capable of night and adverse weather navigation from the operating base to the target and would also allow maximum range ferry flight and continental US (CONUS) operation using conventional radio navigation facilities. Based on Air Force and contractor studies, the estimated unit cost for the A-X was \$1–1.2M (in 1970) depending on purchase quantities, with research and development costs estimated at $240M.

With all these requirements for the concept in hand, the Air Force program office in March 1967 released a RFP for system design studies. Four-month contracts were awarded to McDonnell Douglas, Northrop, Grumman, and General Dynamics in May of that year. The program office also created two government vehicle configurations of its own. The contractor studies were completed by September 1967 and included point designs, supporting data, and detailed development plans. These contractor studies were to be used, along with the government configuration studies, to develop a consolidated concept formulation package (CFP).

7.3.2.3 Questions to the Reader

What concept trade studies would you examine as part of the CFP?

What historical data would you analyze, to convince leadership to fund the development and production of the A-X system?

How would you compare different concepts ... or solutions? What characteristics do you think are important? How would you actually measure them?

7.3.2.4 Final A-X Concept

A CFP was prepared to justify approval for engineering and systems development of this new specialized CAS aircraft (A-X). The CFP defined the CAS mission as having three tasks:

1. Close support fire
2. Armed escort
3. Armed reconnaissance

The first two were considered complementary and the most important of the three tasks. The CFP identified four key characteristics for the CAS mission, by which the final system solutions would be measured:

1. Responsiveness
2. Lethality
3. Survivability
4. Simplicity

Counter to some proponents within the Air Force, responsiveness was not determined by speed, but from the ability to operate from forward area basing and by having extensive loiter time in the battlefield area. Responsiveness dictated that the solution be able to interface with Air Force and Army Command, Control and

Communication (C³) equipment. Extensive analysis was done on how the CAS mission would originate and be conducted. It was assumed that target acquisition would be made by Army units and coordinated through the Army/Air Force communications network. A typical mission scenario consisted of 5 min warm-up and takeoff; climb to 5000 ft at optimum power setting; cruise to 200 NM at 250 knots; loiter for 2 h at 5000 ft; descend to sea level for 15 min of combat at 250–300 knots; climb to 5000 ft at optimum power setting; cruise back to base at 5000 ft and a speed of 250 knots; and descend and land with reserve fuel for 20 min loiter at sea level. Weapon loading for the design mission included seven MK-117 general purpose bombs at 830 lbs each and 1000 rounds of ammunition.

Responsiveness was translated into requirements for combat radius, minimum takeoff distance (driven by the need to operate out of forward bases with short runways), cruise speed, and loiter time. Loiter time and sortie rate were used to determine the size of the force required (including amount of ordnance) to maintain continuous alert over the battle area.

Lethality would be determined by a varied payload of bombs, rockets, guided missiles, and a "new large-caliber high-velocity, high-rate-of-fire gun." Survivability would require protection from small arms, 7.62 mm and 14.5 mm machine guns, antiaircraft artillery (principally the Soviet ZSU-23 mm system), and other surface-to-air missiles.

Lethality studies, in addition to addressing weapon types and loadouts, analyzed the accuracy of tracking and delivery as it related to speed, dive angles, and airframe stability. The A-X needed to be capable of stable deliveries in the region of 200–300 knots and at dive angles up to 50°–70°. This would allow first past delivery with circular error probability (CEP) of 100 ft and a CEP of 50 ft under multiple pass conditions. Simulations with pilots using a ground simulator demonstrated the relation between delivery accuracy and common stability derivatives used in aircraft control system design.

Survivability would depend on maneuverability, redundancy and shielding of critical subsystems (and the pilot), small aircraft size, shielding of infrared sources (such as engine exhaust), and weapon delivery systems. Finally, it was intended that simplicity of design would lead to a shorter development time, lower life cycle cost, reduced maintenance times, increased sortie rates, and the ability to operate from austere bases. Historical analysis of ground fire attrition in WWII, Korea and Vietnam was used to determine which aircraft equipment was most vulnerable and/or likely to lead to loss of an aircraft due to ground fire. The known causes included engine, controls, structure, pilot, and fire. The concept package emphasized designs that could reduce the loss rate. These design features were

1. Fuel could be protected from fire and kept from ignition sources.
2. Manual controls could be made practically invulnerable.

3. Crew compartments could be sufficiently shielded and armored to make pilot losses insignificant.
4. Engines could be shielded, fire protected, and made almost fully redundant. Their oil supplies could be protected.

Maneuverability was also identified as a component of survivability, and the effects of speed and maneuverability were analyzed against probability of aircraft loss for a range of delivery profiles and threat systems. The performance requirements most important for short-range attack were low cruise speed with combat loads, and both high instantaneous and sustained g-limits for initiation and execution of short radius turns without losing altitude. Superior low-speed maneuvering and dive capabilities were shown to enhance close-in fast reattack tactics, allowing operation in visibility half or less than that required for high-speed jet aircraft. High-speed jet aircraft required minimums of 2000 ft cloud ceiling and 3 mile visibility for safe operations, while the A-X was expected to operate with minimums of 1000 ft and only 1 mile.

Maintenance man-hours per flying hours emerged as a key metric and direct indicator of aircraft complexity (simplicity) and were plotted against peak and sustained sortie rates for a range of aircraft operating in Southeast Asia. It would turn out that the most valuable aspect of simplicity was shown to be the ability to operate from austere forward bases, with great improvements in maintenance response time.

Twelve companies were selected to receive the RFP upon its release in May 1970, and six companies responded with proposals. The responders were Fairchild Hiller, Boeing, Northrop, Cessna, General Dynamics, and Lockheed Aircraft. Two winning contractors, Fairchild Hiller and Northrop, were announced publicly on December 18, 1970. Northrop and Fairchild Hiller each signed firm fixed price contracts to provide two prototypes each for a competitive fly-off (see Figure 7.3).

Figure 7.3   Competing prototypes to provide the CAS concept. YA-9A (Northrop) shown on left and YA-10A (Fairchild) shown on right. (Photo courtesy of USAF.)

The winning prototype from Fairchild, the YA-10, would later be integrated with a separate CAS Gun System (GAU-8/A) and become the A-10, which continues successful operation in the Air Force today.

7.3.2.5 Learning Principle

The system concept and preliminary design must follow, not precede, mission analysis.

Developing an effective system concept and architecture is one of the most important responsibilities of the systems engineer in the early life cycle. The A-10 (originally the A-X) concept formulation was rigorous and very traceable to user needs. While the political environment was fraught with service positioning, the early concept trade studies and mission analysis was thorough and resulted in a successful set of system requirements.

7.3.3 System and Subsystem Detailed Design and Implementation Concept Domain

Any major change to a subsystem configuration after preliminary design should be handled swiftly with stakeholder involvement across the system. Also, major changes to a subsystem configuration after preliminary design can have cascading effects across the other subsystems. During this phase of a systems life cycle, a variety of technical and management challenges often are presented to the systems engineer. This section examines the B-2 bomber (Griffin and Kinnu, 2005) and the technical challenges that can be found on any new aircraft or complex system.

7.3.3.1 Early Stealth Research

The concept of stealth technology intrigued aircraft designers from the beginning of the invention of radar. Engineers and scientists investigated various techniques to avoid detection dating to the 1940s. Designs to reduce noise even preceded these efforts. By the mid-1960s, defense laboratories had funded studies to develop radar cross section (RCS) prediction computer programs, radar absorbing materials (RAMs), and shaping techniques. These yielded concepts such as "iron paint," a technique to embed ferrite particles in a quarter inch thick flexible rubberized film. These studies continued into the early 1970s with methods to control the radar returns for representative aircraft shapes. Defense contractors had successfully developed RCS prediction codes, experimented with and flown low RCS drones and had developed analytical techniques for the design of leading-edge RAM.

This early work by both the government and industry resulted in proving low observables could successfully be applied to aircraft, ships, and other vehicles. The current inventory of stealth platforms throughout the military can all trace their roots to these early technology maturation programs and the initial prototype aircraft developed throughout the 1970s. These experiments would eventually lead to the B-2 stealth bomber.

7.3.3.2 B-2 Program Beginnings

The genesis of the B-2 stealth bomber (Figure 7.4) program was a funded concept study initiated by the Air Force in January 1980. This effort was to examine the feasibility of developing a long-range strategic aircraft employing low observables or stealth features. The F-117 stealth fighter development contract had been awarded in November 1978, so the concept for operational deployment of stealth aircraft was evolving and maturing.

Throughout 1980, the program evolved rapidly from a concept study to a full-blown design and proposal effort for the advanced technology bomber (ATB). Due to the effort on the parts of the government and industry during this time frame, the requirements for the B-2 were derived, traded, balanced, approved, and documented. Each contractor had assembled design teams and performed many SE analyses and trade studies to define and refine their initial system conceptual approach.

Figure 7.4   B-2. (Photo courtesy of USAF.)

The primary role of the ATB was for long-range, high-altitude cruise penetration of the Soviet radar network. There were 14 missions stipulated for use in calculation of range, payload, structural durability, and design spectrums. The proposed system concepts were predominantly designed for high-altitude missions. There was a mission for low-altitude operation stipulated in the RFP, but this was a fall-out capability, and not to be used as a specific design point. The evaluation of the contractors' proposals proceeded throughout December 1980 and into February 1981. It was then that the government concluded that growth provisions for low-altitude capability would be a prudent hedge against an ever-changing and maturing radar threat. Accordingly, a modification request to the RFP was issued in April 1981 to request a study for the impact on the design to include a significant low-altitude penetration capability. The scope of the request was to examine completely new designs, in addition to studying a modification to the current proposals. Only hindsight would be able to tell the implications of this change.

7.3.3.3 Questions to the Reader

What would you do to effectively handle the low observability (i.e., stealth) requirements throughout the design, development, and production?

Is low observability like other SE 'ilities', such as reliability, availability, and maintainability?

What could be some technical risks in adding the low-altitude mission to the current design?

7.3.3.4 Contract Award to Critical Design

Northrop was announced as the winner in November 1981 and the $9.4B full-scale engineering and development contract was signed on December 4, 1981. The original contract schedule had a first preliminary design review (PDR) for several subsystems after 11 months, a second PDR for the entire system after 28 months, and a critical design review at 48 months. The "first flight" of a developmental B-2 was planned for 72 months after contract award.

The first milestone after the first PDR was "configuration freeze," scheduled 9 months prior to the full PDR. It was at this contractor-imposed milestone that the need for a major redesign was discovered. The redesign would be the largest single internal event that occurred during development and contributed both to the slip in the first flight date and to the cost increase of the program.

The analysis of wing design performance to the low-altitude mission requirements became available in January 1983. At this time, the air vehicle team advised the Northrop program manager that the analysis revealed a substantial wing bending moment increase over earlier estimates due to more severe aeroelastic effects incurred at high-speed, low-altitude flight conditions. This would result in an inability of the aircraft controls to provide any meaningful gust load alleviation.

The analysis also revealed that distribution of the loads across the centerline was strongly asymmetric with the forward weapon bay bulkhead receiving 70% of the total, leaving only 30% through the rear bulkhead. The combination of increased bending movement and poor distribution led to high internal loads in the inboard forward wing structure.

7.3.3.5 Baseline Configuration Change

Based on the serious nature of the problem and faced with the likelihood of significant revision to the baseline configuration, the program manager ordered a number of actions:

1. The flight controls, aerodynamics, and the configuration teams were directed to study alternative control configurations that would provide assured controllability and eliminate the adverse aeroelastic response.
2. A structures design team with members from all three airframe contractors was tasked with reviewing the structural arrangement and proposing alternative approaches to alleviate the high internal loads.
3. All program elements were tasked to review all existing concerns and potential alternatives to ensure that potential future change requirements were not overlooked.

The response to these directions was a multipronged examination by the SE task teams, with support from other teams, to identify all existing problems and generate and evaluate potential solutions. This 6 month redesign resulted in a program impact of $1.5–2.5B (rough order of magnitude estimate at that time). However, the proposed configuration and structural changes provided a viable solution to the aircraft design, as it was determined that the original design would not have met the original specification without them. A partial list of changes included

1. A control system redesign that could simultaneously handle the gust response of the aircraft and reduce the wing bending moment impact of gusts.
2. A more balanced bending load distribution on the wing (60/40 in lieu of 70/30).
3. A significant reduction in local internal loads.
4. A significant reduction in the susceptibility of the structure design to fatigue failure during the life of the aircraft.
5. An acceptable solution to another identified problem, the flow and low observability of the engine inlets.
6. Improvements in propulsion efficiency and drag reduction.

The B-2 critical design review (CDR) was held per the contract schedule (December 1985), but as was expected, the subsystem designs were only 20% complete. Overall, the resulting change in target cost was estimated to exceed $2.3–2.9B. The schedule accommodated the concerns with regard to the status of the subsystem definitions, procurements, and design development. The first flight event was scheduled to occur in December 1987 as an original contract milestone. As a result of the configuration change, the program office had assessed at CDR that this milestone would occur in December 1988. The actual flight date was 6 months later (July 1989).

7.3.3.6 Learning Principle

Major configuration changes after preliminary design can have huge cascading effects (design, schedule, and cost) across the other subsystems. Handle these technical problems impacting system configuration swiftly.

The identification of a major aeronautical control inadequacy of the baseline configuration just 4 months prior to the formal configuration freeze milestone caused an immediate refocus to develop a substantially revised design. While the program response to the crisis was rapid and effective, the magnitude of the impact on the downstream cost and schedule was not anticipated nor predicted. The effect of the reconfiguration on the maturity of all the other air vehicle subsystems (flight control, environmental control, electrical, landing gear, etc.) was far greater than projected. The subsystems were mostly vendor-supplied equipments and most were in negotiation to the original baseline at the time of the reconfiguration. After the new configuration was determined, the requirements for the subsystems changed to such a degree that the equipments had to be modified. It took longer than anticipated to recognize the growing problem of getting all the specifications updated and to identify the lagging equipment maturity that resulted. Thus, the reconfiguration required a second iteration of the design requirements and their flow down to the many suppliers and their detailed designs.

7.3.4 Systems and Interface Integration Concept Domain

Perhaps, the first thing that comes to mind when one reads "systems and interface integration" is unit testing. Major DoD systems acquisition programs typically employ the equivalent of what is generically referred to as a system integration laboratory or SIL for short. SILs are used to confirm the proper functioning of interfaces between various system components. Although testing is important, systems and interface integration includes more.

NASA's SE Handbook (NASA, 2007) states

> The management and control of interfaces is crucial to successful programs or projects. Interface management is a process to assist in controlling product development when efforts are divided among parties (e.g. Government, contractors, geographically diverse technical teams, etc.) and/or to define and maintain compliance among the products that must interoperate.

In their original paper outlining their framework for SE case study research, Friedman and Sage (2004) note the following:

> In general, component and subsystem testing proved to be insufficient in predicting problems that are likely to occur in the many stages of systems integration. It is necessary to insure that systems integration issues are dealt with throughout the lifecycle …

The following vignette proceeds from this view of systems and interface integration.

7.3.4.1 Cold War Context

Relations between the United States and the Soviet Union soured immediately after the end of WWII, leading to the decades long Cold War. As the Cold War escalated with each side desperately trying to outpace the other in terms of military power, the Peacekeeper Intercontinental Ballistic Missile (ICBM) was conceived to close a perceived gap in strategic nuclear capability. Building on the technology matured during the earlier minuteman program, advancements were to be made in accuracy, lethality, and survivability. Survivability spawned the most publically debated and hotly contested aspect of the program—namely, where and how to base the missiles.

Peacekeeper was an important U.S. weapon system contributing to the mutual deterrence with the Soviet Union, ending with their collapse in 1991 (Stockman and Fornell, 2008). The system was never used and the Cold War came to an end.

7.3.4.2 Peacekeeper ICBM Program

Peacekeeper began in 1973 as the "Missile eXperimental" (MX) program (see Figure 7.5). Basic technologies were leveraged from the previous minuteman program, such as the propulsion system. A new warhead design and an advanced inertial measurement unit were selected for improved accuracy and higher yield. Martin Marietta was awarded a prime contract for full-scale development in 1978 and a team of 12 associate contractors began their work to reach IOC in 1986.

The four propulsion systems used on the missile were each produced by a different contractor. The Department of Energy developed the warhead. The guidance and control system, reentry vehicle system, launch canisters, launch control system, basing support equipment, and missile transporter, ground equipment, and flight test support equipment were the responsibility of other associated contractors. Many points of interconnection between the pieces provided by this group of associate contractors meant that a large number of interfaces would need to be controlled. Achieving clear, fact-based technical communication and unity of effort with this diverse group was critical to achieving program performance objectives and to meeting the schedule. Controlling and coordinating changes was also

Figure 7.5 Peacekeeper launch (Photo courtesy of USAF.)

critical, particularly given the continuing debate around the choice of a basing mode.

The basing mode was one key system interface that proved to be problematic. From the outset, the choice of a basing was embroiled in political turmoil. Congress in 1976 refused to fund Peacekeeper using existing minuteman silos because they believed them to be vulnerable to a first strike. More than 100 basing variations were discussed and in 1980, the office of technology assessment (OTA) looked at 11 options in more detail. OTA, shut down in 1995, was a nonpartisan analytical agency established to assist Congress. The OTA report contained the initial basing plan chosen by President Carter in 1979 referred to as mobile protective shelters (MPS). This plan called for the shuttling of missile launchers and "dummies" between 4600 shelters. The shelters would be scattered across vast areas of public land throughout multiple states creating environmental concerns. These concerns contributed to MPS cancellation in 1981. In 1982, President Reagan pushed the "dense pack" concept wherein the missiles would be deployed in closely spaced silos. Congress again cut funding for the program. In 1983, the Reagan administration then devised a plan to put 100 missiles in silos while development continued on other basing options. Congress cut the plan to 50 missiles.

7.3.4.3 Questions to the Reader

How could the indecision about the basing mode affect the Peacekeeper design? How would you stabilize and integrate the work of the twelve associate contractors? What could you do to minimize design risk presented by the lack of a defined basing mode?

7.3.4.4 Peacekeeper Program Progress

Despite the turmoil created by the lengthy basing debate, the development of the missile progressed relatively smoothly. The Air Force retained the role of overall weapon system integrator and "owned" the SE process, thus providing for unity of effort across the 12 associate contractors. In that role, the government had control of the system specification, associate contractor specifications, and interface control documents between associate contractors. Process discipline was put in place through the implementation of the systems requirements analysis (SRA) process that had evolved during the Minuteman program. Associate contractors were contractually required to implement the SRA process.

The SRA process contained four key elements:

1. The operational requirements analysis defined and allocated the technical requirements for the prime operational equipment for each associate contractor. It also included tech orders, training, and related activities and described the requirements for the operational personnel.
2. Test planning analysis included the requirements and plans for testing. This included the integrated test plan by which all developmental and operational testing was planned leading up to and including flight testing. This also developed the requirements for test facilities.
3. Logistics support analysis developed the requirements for sustainability and for deliverable test equipment for both the depots and the field sites.
4. Assembly and check-out (A&CO) analysis developed the A&CO requirements on the prime equipment as well as special equipment to accomplish A&CO in the field.

The SRA process was guided by a series of forms that captured end-item requirements as well as personnel requirements. Perhaps, the most important one was simply called the Form B—the requirements analysis sheet. The data on the Form Bs became both the historical record showing the genesis of each requirement and the basis for the specifications. The SRA process was maintained as well as practiced over the life of the system. SRA provided the necessary basis for clear, fact-based, technical communication, especially across interfaces.

7.3.4.5 Learning Principle

The systems integrator must maintain a rigorous systems engineering process.

System and interface integration goes well beyond testing and includes integration of the work efforts of various associate contractors. The formal SRA process brought significant discipline to the SE process followed by the associate contractors. While at times tedious, it constrained the contractors in adding capabilities (and making changes) and enforced engineering discipline.

One of the benefits of an enforced SE process was that it required analytical decision making. This is difficult in the absence of complete and accurate data and the presence of freely given expert opinions. Fortunately, extensive technical and programmatic data was collected and retained both during development and operational usage. Therefore, most decisions were based, to a large extent, on fact rather than hypothesis. A former senior program official commented (Stockman and Fornell, 2008):

> It [SRA] provided the structure that saw us through despite the enormous political upheaval and the challenges of managing more than ten associate contractors to deliver equipment close to schedule that effectively operate together so that a 200,000 pound missile could launch within about one minute of the command and deliver 10 warheads over 5000 miles away in about 30 minutes with each hitting different targets with a precision of a few hundred feet.

7.3.5 Validation and Verification Concept Domain

Many SE texts state that verification answers the question, "Did you build it right?" This statement refers to assessing product performance against the specification. Validation, on the other hand, answers the question, "Did you build the right thing?" In this case, the product is assessed against the intended operational environment with anticipated operators and users. Validation and verification are tools by which factual data is obtained to answer the above questions. One aspect which affects the time of when factual data becomes available is program concurrency. This relates to how much overlap there is between development, production, and test. Timing of factual data can present serious cost and schedule consequences.

7.3.5.1 C-5 Galaxy Acquisition

A behemoth compared to most aircraft, the C-5's size was driven by the need to move Army divisions and their bulky "outsized" cargo (i.e., cargo that exceeds 1000 in. L × 117 in. W × 105 in. H). See Figure 7.6. In fact, the Army had asked

Figure 7.6   C-5. (Photo courtesy of USAF.)

the Air Force to have the capability of airlifting a tactical airborne assault force of two and two-thirds divisions, plus one additional division, to potential combat theaters worldwide (Launius and Dvorscak, 2001). Early requirements for an aircraft capable of carrying outsized cargo were documented in 1964 in a specific operational requirement document for the Cargo eXperiment–Heavy Logistics Systems (CX-HLS). Boeing, Lockheed, and Douglas received study contracts to accomplish preliminary conceptual studies and delivery of preliminary designs. After these studies, Secretary of Defense Robert S. McNamara announced in December 1964 the decision to proceed with procurement of the C-5 (Griffin, 2005). These three contractors intensely competed for this contract, and Lockheed announced the winner in September 1965. Key factors of the acquisition strategy that contributed to program difficulties included high levels of concurrency, the total package procurement concept (TPPC), and a weight empty contract guarantee.

TPPC was a new concept for weapons system procurement devised by the Honorable Robert H. Charles, assistant secretary of the Air Force for installations and logistics. The purpose was to enhance the competitive environment and to create incentives for cost-efficient programs that would reward contractors for controlling development and production costs. Secretary Charles had previously worked for McDonnell Aircraft Corporation and had a "… first-hand view of the problems created by the lack of competition that existed under the contractual arrangements employed on weapons acquisition programs." During the 1950s, contractors commonly submitted bids with an optimistically low price for development, but the average cost overrun by the time development and production were complete was 320%. During the early 1960s, many initiatives were discussed for increasing competition and controlling costs. Almost all the changes had an objective of better defining the total system cost, not just the development cost. The introduction of life cycle cost—a vital prerequisite for any TPPC—was one of the lasting outcomes of this reform era.

In October 1965, Lockheed was awarded a firm fixed price incentive fee contract for the development of the system and a first production quantity of 58 C-5A aircraft. The contract also included an option for a second quantity purchase of 57 aircraft. The price for the second quantity of aircraft was based on the price paid for the first quantity of aircraft, but a cost acceleration clause allowed the contractor to recoup losses as the number of aircraft in the second quantity increased. There was also an option for a third production lot of 100 aircraft, but it was never exercised. The contract also included a guaranteed weight empty requirement that called for substantial financial penalties for exceeding the government's predetermined weight estimate. In essence, TPPC in combination with the weight empty guarantee forced the contractor to answer the question "how much will this weigh" before the design was complete, before the aircraft had been built, and before completion of any testing to confirm the validity of weight estimates of subsystems or component parts.

7.3.5.2 Guaranteed Empty Weight Requirement

During source selection leading up to the contract award in October 1965, the Air Force notified all competitors about various proposal deficiencies that needed to be corrected. They were given only 4 days to correct and resubmit their proposals. After months of work developing initial design proposals that accurately described aircraft performance and specifications, each competitor was forced to very quickly make serious changes. Lockheed, the ultimate winner, made considerable changes to the wing, leading-edge flaps, engine thrust reversers, and inlets. All of the analysis, testing, and data that had gone into the original proposal could not be repeated in 4 days. This rush to award a contract contributed to a contractual guaranteed weight empty that was unattainable.

Problems in meeting the guaranteed weight empty requirement ultimately resulted from the approach Lockheed used to produce its weight estimates for the initial design proposal, the revised proposal guaranteed weight empty, and the expected weight growth. Lockheed management applied a 5% reduction factor to the originally developed weight empty to arrive at the original proposal value of £302,495. This factor was extrapolated from smaller aircraft and assumed technological advances that would lead to future weight savings. Lockheed adjusted the new guaranteed weight empty to a value of £318,469 to account for the unknown weights of the very quick redesign. This figure was based on an estimate for the new wing area structure and flap/slat changes. Lockheed's decision to include the weight of the redesign in the £318,469 estimate would plague the development of the C-5 for the remainder of the design program.

The expected weight growth used in determining the target weight of the aircraft was allocated on the basis of experience with the C-141. However, these allocations were fundamentally flawed because they were based on a skewed guaranteed weight empty.

7.3.5.3 Questions to the Reader

What do you see as the ramifications for validation and verification efforts resulting from the guaranteed aircraft empty weight?

What control mechanisms would you employ as part of the validation and verification process to ensure that the aircraft design was not overly dominated by the guaranteed weight empty requirement?

7.3.5.4 Weight Problems Continue

The goal of the target weight calculation was for the air vehicle to be initially designed to a weight level sufficiently low to permit reasonable weight growth without exceeding guaranteed weight at the time of delivery. Weight growth allowances of 4% of the controllable weight, 1.5% for manufacturing variations prior to first flight, and 5% for basic vehicle weight growth account for an initial target weight that is approximately 5% below operation weight. In guaranteed weight empty terms, this initial target weight was over £2000 less than that listed in the original proposal. If the aircraft had not been redesigned, this methodology would have been a logical way to meet the goal. However, the redesigned wing, coupled with aggressive weight goals, made meeting the guaranteed weight empty goal almost impossible. The government program office recognized the problem in mid-1966 and alerted Lockheed. Lockheed's request that the government increase the guaranteed weight empty requirement by £14,000 was denied. The Air Force also denied requests to pursue other possible solutions in meeting requirements such as increasing engine thrust. The Air Force had no intention of making any design trade-offs pertaining to weight and performance. Lockheed began an extensive, ambitious, and overly aggressive three-phase program of monitoring, controlling, and reducing aircraft weight.

The first phase would continue to the completion of the preliminary design. The second phase continued from that point to the first flight. The third phase was to end with delivery of the first fully compliant C-5.

During Phase I, the target weight allocations were broken down and distributed to the many functional design groups. Lockheed developed procedures to use the target weights and allocations as a control over the release of job packages for the design: no jobs were to be released that were over the target weight. The efforts to reduce weight became more challenging when further detailed analysis dictated configuration changes that drove revisions to the weight allocations. Design groups found themselves having to do more with less allowance for weight.

Lockheed had programs to encourage innovative weight reduction ideas among employees. The first program lacked incentives, but still managed to gather over 1700 ideas, of which 64 were incorporated for a weight savings of approximately

£4000. This program was replaced at 50% design release with a program that gave $200 per pound to employees with weight savings ideas that were eventually incorporated into the aircraft.

Other mechanisms to restrict and reduce weight during Phase I included controlling the nominal thickness of aluminum sheet metal, chemical milling of parts to remove excess material, and monitoring the thickness of paint at application. Lockheed also developed programs for contract vendors to meet weight target allocations or face penalties for overweight deliveries. These and the many other aggressive weight reduction programs still fell short of enabling Lockheed to meet its target at 50% release.

Phase II was implemented during production at a time when it became clear that extreme measures had to be taken to meet guaranteed weight empty. A report led to the "Save Cost and Limit Pounds" program, called SCALP, which combined the weight and cost relationship to evaluate parts considered for weight reduction. SCALP guidelines mandated that changes of £5 or less had to cost less than $40 per pound to implement. Changes of more than £5 had to cost less than $75. Special priority was given to changes that saved cost and weight. High cost and weight areas of the aircraft were reviewed to further the weight reduction efforts in an affordable manner. Because this program was initiated at the completion of the stress analysis, it was applied to many over-strengthened areas for a large weight savings.

The final phase of the weight reduction efforts carried over many of the same programs from both Phases I and II. Phase III occurred during flight test; therefore, any changes with large manufacturing impacts were not considered programmatically feasible.

The weight reduction efforts were so concentrated on conserving cost that they compromised the ability to meet other requirements. The negative consequences of the weight reduction efforts for the "over-strengthened" areas became very apparent during verification testing after the 1969 static test failure of the wing at 125% of the design limit load. A wing that had been thoroughly and meticulously designed for the original proposal had borne a large burden of the weight reduction efforts on a heavier aircraft. From the estimated weight report, in the time frame starting with the 60% design release point and proceeding to the actual weight of the ninth aircraft, the overall weight of the wing was reduced by almost £4000 or about 4.5%. At one point in the wing's development, it had reached a low weight of £78,100, 8% below the weight at proposal. The progression of the wing weight can be seen in Figure 7.7.

The overall aircraft saw the lowest estimated empty weight of £310,000 in July 1966 before the impacts of the redesign were fully understood. The empty weight then immediately increased to almost £323,000 before Lockheed again tried to reduce it.

After the second static test failure in March 1971, payload and maneuvers used throughout the operational fleet had to be limited. Widespread fatigue cracks

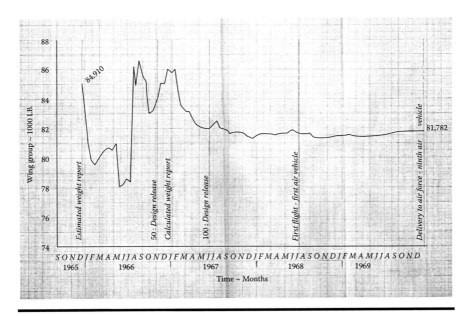

Figure 7.7   Actual wing weight data for C-5A through the ninth delivered aircraft.

appearing in the test articles and in the test aircraft forced the program office to issue restrictions for maneuvers, payload, operations in turbulence, and ground–air–ground cycles. The new limits on operational usage yielded a service life prediction of only 8,000 h, far short of the required 30,000 h of life at a more severe usage spectrum.

A new wing design in the late 1970s would remedy this problem. The completed wing redesign and retrofit brought the service life of the C-5A up to the design specifications and the aircraft weight to £332,500. Ironically enough, this weight value is less than 5%, or about £14,000, over the original empty weight estimate of £318,469 given in the second proposal.

7.3.5.5 Learning Principle

The usefulness of validation and verification data as a tool is only as good as the ability to react to the consequences of its analysis.

Aircraft weight issues in the C-5 program underscore the importance and the implications of decisions that, at the time, seem to have small consequences. Verification testing highlighted problems with strength and fatigue life of the wing, but these results came too late to influence the design or fabrication of initial production aircraft. As a result, initial aircraft operations were conducted with flight restrictions and a costly wing replacement program was needed to meet original service life requirements.

7.3.6 System Deployment and Post-Deployment Concept Domain

Prominent war theorist General Carl von Clausewitz wrote about the fog and friction of war in his seminal work *On War*. In war, execution does not go exactly as planned amid much uncertainty. Likewise, deployment of a system or product may not go exactly as planned when first placed in the hands of those who operate and maintain them. There are many pressures for change. Operators find new ways to use the system while gaining familiarity with its strengths and weaknesses. New capabilities must be added to meet new needs not envisioned at the time of first deployment. Costly military systems are kept in use over many years or even decades. Over time, the system begins to wear out. Parts break down, malfunctions become more frequent, and supplier sources diminish. Maintenance costs grow. Eventually, the decision to continue operating the legacy system, make major upgrades, or to replace the system must be made. The SE process must address these pressures for change.

7.3.6.1 A-10 Aircraft for Close Air Support

As noted in a previous vignette, CAS is air action by fixed and rotary wing aircraft against hostile targets that are in close proximity to friendly forces and which require detailed integration of each air mission with the fire and movement of those forces. The A-10 Thunderbolt II design and construction was optimized for performance in that specific operating environment (Jacques and Strouble, 2008). Proving its mettle since the 1970s in Operations Desert Storm, Allied Force, Enduring Freedom, and Iraqi Freedom, the A-10 is projected to remain in the operational inventory until 2028 (see Figure 7.8).

7.3.6.2 Debate Concerning CAS Aircraft

Just as concern over Army helicopter developments contributed to the early pressures that led to the A-X concept studies, so too did pressure from the Army in 1982 again drive a debate over CAS aircraft. The Army unveiled its new AirLand Battle Doctrine that envisioned a much faster and free-flowing battle without a traditional battle line (Jacques and Strouble, 2008). Firepower and maneuverability would slow enemy advance and simultaneously attack their reserve troops before they could be used to engage friendly forces. Enemy troops would be forced to either retreat or surrender. Combat of this nature favored a faster aircraft and required less reliance on direct coordination with ground troops. Also in this timeframe, the Army had been given approval to begin production of the AH-64 Apache helicopter that was to provide the organic CAS they had long sought.

As production of the A-10 was completed in the early 1980s, Air Force leaders began to question whether or not the A-10 was the CAS aircraft needed for the

Figure 7.8   A-10. (Photo courtesy of USAF.)

future. After several studies, a 1985 report projected that the A-10 would lose its effectiveness in mid- to high-intensity conflicts by the mid-1990s. The Air Force held the view that a modification of an existing aircraft that could be fielded in the mid-1990s was considered necessary to avoid competition with the top new fighter development priority; the advanced tactical fighter—precursor to the F-22. In December 1986, the Air Force recommended to the OSD that the A-10 be replaced by a modified variant of the F-16 fighter.

This request by the Air Force was disapproved. A protracted, highly political debate ensued with Congress authorizing funds to provide a CAS master plan by the end of 1989. World events would delay implementation of that plan. Iraq had invaded Kuwait and President Bush had ordered U.S. forces to the Middle East. The final decision in the fall of 1990 called for the Air Force to retain two wings of A-10s and to retrofit up to four wings of F-16s to perform the CAS and related battlefield air interdiction missions.

7.3.6.3 Desert Storm Events

Desert Storm derailed the modified F-16 plan and set the stage for retention of the A-10. The Government Accountability Office reported that the A-10 had the highest sortie rate, with an average of 1.4 sorties per aircraft per day, and delivered more guided munitions (almost 5000 Maverick missiles) than any other aircraft type. An Iraqi regimental commander described the A-10 as "the single most recognizable and feared aircraft," noting its ability to conduct multiple raids per day, loiter around the battlefield, and attack with deadly accuracy. The A-10 survivability was also generally confirmed (Jacques and Strouble, 2008): "… the Hog's redundant flight control system allowed crippled planes to fly home. Aircraft battle damage repair (ABDR) crews repaired in-theater all but one of the estimated seventy damaged A-10s during this war— and of those, twenty suffered significant damage. These repairs were usually quick, and used cheap, accessible materials."

Twenty-four F-16A/B aircraft were converted to a CAS configuration prior to deployment in Desert Storm. Configured with a pod-mounted GAU-13/A four-barrel gun and a laser target acquisition system, these F-16s did not perform well in Desert Storm. The pylon mounted gun shook the aircraft, which made it hard to control the round placement. The higher speed of the F-16 did not provide enough time when approaching a target for gun engagements. After several days of operations, the gun pods were removed and the "CAS" F-16s returned to more standard F-16 operations.

7.3.6.4 Questions to the Reader

What factors limit the useful service life of an aircraft?
What technical information is needed to determine the viability of retaining the A-10 for the CAS role?
Would you recommend physical inspection of fielded aircraft before making a decision about retention? If, so why?

7.3.6.5 Structural Integrity and Retention of the Warthog

After the war in 1992, the Air Force made a decision to retain the A-10, although in reduced numbers. It faced the question of how to extend the viable life of these aircraft, which were now a decade old and some of which had been stressed in combat (Jacques and Strouble, 2008). The answer to that question depended on how well the aircraft had been maintained and the actual physical condition of each aircraft.

Critical to the service life of any aircraft is maintaining structural integrity. Initiated in 1958, the Aircraft Structural Integrity Program (ASIP) was established to monitor and evaluate the structural health of all Air Force aircraft throughout their operational life. The objective was to discover and correct problems before the occurrence of a structural failure attributable to fatigue. Air Force directives required a plan for monitoring and evaluating the structural health of each type of aircraft in the operational inventory.

ASIP calls for a damage tolerance assessment (DTA) and a force structural maintenance plan (FSMP). The initial DTA for the A-10 was done by Fairchild in 1980 with several subsequent reassessments. Grumman Aerospace, who later merged with Northrop to become Northrop Grumman, took over the A-10 program from Fairchild Republic in 1987 and delivered an updated DTA in 1993. The DTA analyzed 52 control points in the wing, and comparison between the 1980 and 1993 DTAs indicated that service lives were reduced on 8 of these control points. The 1993, FSMP was intended to influence procedures for inspection, repairs, and modifications and established inspection intervals based on service life and safety limits. It was intended to be accomplished as programmed inspections on all aircraft.

However, the FSMP inspection requirements were not incorporated into the inspection and maintenance technical orders for the aircraft. Technical orders are military orders that maintenance personnel must comply with or they face potential disciplinary action. Inspections were accomplished using random sampling as opposed to monitoring all aircraft.

Further breakdown in the implementation of an adequate FSMP resulted from the loss of experienced personnel due to the base realignment and closure (BRAC) decision in 1995. This decision to save DoD costs closed McClellan AFB, Texas, and consolidated the maintenance and repair operations to Hill AFB, Utah. By the time the BRAC decision was fully implemented in 2000, 80% of the experienced workforce had been lost. Compounding the situation, in 1997, Lockheed Martin Systems Integration (LMSI), formerly IBM Federal Systems Division in Owego, New York, was awarded an indefinite delivery/indefinite quantity contract to take over as the new prime for A-10. It should be noted that LMSI was not an aircraft company and did not have aircraft repair infrastructure. Note: the plan was that Lockheed Martin would merge with Northrop Grumman and thus gain the necessary expertise for A-10 aircraft maintenance; unfortunately, this merger was not approved.

Inspections conducted in 1995–1996 discovered cracks in several wing locations due to fatigue. Most of the cracks were consistent with crack growth curves updated in 1993. However, two cracks at wing station 23 (WS 23) were underpredicted and one of the cracks was of "near-critical" size. Classified as minor cracks, the implementation of the FSMP was not reconsidered. In 1998, Northrop Grumman was tasked to provide a cost-effective structural enhancement program focusing on the most critical areas. In August 1998, they delivered a report entitled "A-10A Aircraft Wing Center Panel Rework-Fatigue Life Improvement." The plan detailed structural changes required to support a 16,000 h service life. The report recommended immediate implementation and verification using a full-scale fatigue test on a modified wing. This report formed the basis of the subsequent program called "HOG UP" to extend the structural life to the year 2028. Further, it was based on the assumption that the 1993 FSMP had been implemented and did not consider the impact of crack data or new fatigue-sensitive locations that had been identified by field inspections. The HOG UP program was initiated in 1999 as a repair program. Since it was a repair program, "appropriate configuration control concerns, such as technical analysis of service life, technical contents of the program, and method to evaluate an organic or contractor prepared engineering change proposal did not occur."

The HOG UP program eventually expanded. Air Combat Command requested center wing fuel bladder replacement, rework of the flight control system, and nacelle fitting inspections. However, a Red Team independent investigation noted in 2003 that no composite estimate of the risk of structural failure had been generated and expressed concern that the repair might not result in the intended life extension. A further complicating factor had to do with the problematic cracks at WS 23. In 2001, the WS 23 crack was reclassified as critical, and a new technical order was issued for the inspection of the wing center panel and WS 23 fastener holes. Estimates in 2003 were that 35 aircraft would require refurbished wings associated with the WS 23 repair. Although not originally part of HOG UP, the WS 23 inspection and repair was subsequently scheduled to be conducted concurrently with the expanded HOG UP repairs. Full-scale fatigue testing to validate the HOG UP repair had not yet been done, so the Red Team concluded that the actual structural condition of the fleet remained unknown, and the repair was "unvalidated for extending the lives of A-10 wings to 2028 (~16,000 h)."

Subsequent to the Red Team report in February 2003, the wing undergoing full-scale fatigue testing failed short of the 16,000 h life expectancy. Thin-skin wings coming into the depot were failing inspection at an increasingly higher rate. It became clear that the Air Force would run out of serviceable wings by about 2011. By 2005, the failure rate of the thin center panel wings coming in for service life extension was hovering near 30%. In 2005, the AF completed a business case analysis that considered options for structural life extension. Selecting an option to buy 242 wings to replace WS 23 failures, the Air Force prepared in early 2006 a budget justification for production of newly manufactured "thick-skin" wings to replace the remaining "thin-skin" wings in the A-10 inventory. In the end, aerospace engineering spectrum, LLC would be awarded the contract to build a computer model for a wing that would be manufactured by Boeing, the winner of the new wing contract in 2007. The wing would be installed on an aircraft built by Fairchild Republic, for which Lockheed Martin was now the prime; this was the new complicated arrangement of sustainment contractors for the A-10.

7.3.6.6 Second Life for the Modern Day Hog

Coincident with the HOG UP program, several other upgrade programs were undertaken (Jacques and Strouble, 2008). In the early 1990s, the aircraft was modified to incorporate the low altitude safety and targeting enhancements (LASTEs) system. LASTE added ground collision avoidance warnings, an enhanced attitude control function for aircraft stabilization during gunfire, a low-altitude autopilot system, and computed weapon delivery solutions for targeting improvements. Starting in 1999, the A-10 was upgraded with the installation of an embedded global positioning system/inertial navigation system to provide improved navigation and situational awareness. The Precision Engagement (PE) program, which results in the modified aircraft being re-designated as A-10Cs, was awarded to Lockheed Martin in 2005 and includes the addition of enhanced precision target engagement capabilities. A-10Cs are able to carry the INS/GPS guided joint direct attack munitions and the wind corrected munitions dispenser. As of January 2008, the 100th A-10C conversion had been delivered. The PE upgrade is intended to evolve the A-10 from its origins as a cold-war tank killer, to an aircraft capable of performing a wide range of operations to support the global war on terror and other contingencies. The Air Force has committed to sustaining the A-10 for the foreseeable future.

7.3.6.7 Learning Principle

Successful design, development, and production are not enough to sustain a system throughout its life cycle.

The Air Force lost awareness of the structural health of the A-10 fleet. The intended repair proved to be costly and did not provide the required service life. This resulted in the manufacture of new wings for all remaining thin-skin aircraft. Had the Air Force maintained awareness of the structural health of the A-10, the decision to re-wing could have been made earlier, possibly obviating the need for the costly and unnecessary wing components of the HOG UP program.

7.3.7 Life Cycle Support Concept Domain

Life cycle support planning and execution includes the aspects of SE concerned with optimizing total system performance while minimizing total ownership costs. Effective sustainment of systems begins with the design and development of reliable and maintainable systems. In addition, such considerations include configuration management, item identification, technology refresh, and reducing the logistics footprint. Programs structured with adequate life cycle performance as a design driver will be capable of better in-service performance and will be capable of dealing with unplanned, unforeseen events (even usage in unanticipated missions). This vignette will summarize the aspects of designing, developing, launching, and sustaining the Hubble Space Telescope (HST).

7.3.7.1 Hubble Space Telescope

The HST has been the premiere orbiting astronomical observatory; it is already well known as a marvel of science (Mattice, 2005). Viewed with the clarity that only time and hindsight provide, the HST program certainly represents one of the most successful modern human endeavors on any scale of international scope and complexity. Launched in 1990 and scheduled to operate through 2010, HST carries and has carried a wide variety of instruments producing imaging, spectrographic, and photometric data through both pointed and parallel observing system components (Figure 7.9). Over 100,000 observations of more than 20,000 targets have been produced for public retrieval. This enormous volume of astronomical data collected by Hubble has taught us about our universe, our beginnings, and, consequently, about our future. While operationally successful, the acquisition and SE, like on

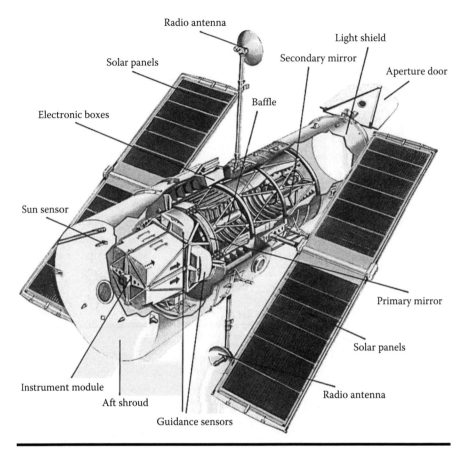

Radio antenna

Light shield

Solar panels

Secondary mirror

Aperture door

Baffle

Electronic boxes

Sun sensor

Primary mirror

Solar panels

Instrument module

Radio antenna

Aft shroud

Guidance sensors

Figure 7.9   Hubble Space Telescope. (Photo courtesy of NASA.)

any large complex system, was not without flaws. Life cycle design considerations could have possibly contributed the most toward the Hubble success.

For decades, astronomers dreamed of placing a telescope in space well above the Earth's atmosphere, a complex filter that poses inherent limitations to optical investigation and observation of celestial bodies. A 1923 concept of an observatory in space was suggested by the German scientist Hermann Oberth. In 1962, and later in 1965 and 1969, studies at the National Academy of Sciences formally recommended the development of a large space telescope as a long-range goal of the emerging U.S. space program.

7.3.7.2 Program Initiation

With the approval of the Space Shuttle program and with the shuttle's inherent capacity for man-rated flight, large payloads, and on-orbit servicing, stability, and control, the concept of a large telescope in space was seen as practical (albeit at significant expense and with major technical and SE challenges). In 1973, NASA selected a team of scientists to establish the basic telescope and instrumentation design and Congress provided initial funding. In 1977, an expanded group of 60 scientists from 38 institutions began to refine the system concept and preliminary design. The use of preprogram trade studies to broadly explore technical concepts and alternatives was essential and provided for a healthy variety of inputs from a variety of contractors and the government (NASA centers). The studies had exercised various concepts of operation for launch, deployment, and servicing (including on-orbit vs. return to earth), with cost trade-offs a major consideration.

NASA formally assigned systems responsibility for design, development, and fabrication of the telescope to the Marshall Space Flight Center in Huntsville, Alabama. Marshall subsequently conducted a formal competition and selected two parallel prime contractors in 1977 to build what became known as the HST. Perkin-Elmer in Danbury, Connecticut, was chosen to develop the optical system and guidance sensors, and Lockheed Martin Space Corporation of Sunnyvale, California, was selected to produce the protective outer shroud and the support systems for the telescope, as well as to integrate and assemble the final product.

The design and development of scientific instrumentation payloads and the ground control mission were assigned to Goddard Space Flight Center in Greenbelt, Maryland. Goddard scientists were selected to develop one instrument, and scientists at the California Institute of Technology, the University of California at San Diego, and the University of Wisconsin were selected to develop three other instruments. The European Space Agency agreed to furnish the solar arrays and one of the scientific instruments.

7.3.7.3 Shuttle Challenger Disaster

Development, fabrication, integration, and assembly of Hubble was a daunting, almost 10 year process. The precision-ground mirror was completed in 1981. Science instrument packages were delivered for testing in 1983. The full-up optical assembly was delivered for integration into the satellite in 1984, and assembly of the entire spacecraft was completed in 1985. The launch was scheduled for October 1986. Unfortunately, the Space Shuttle *Challenger* disaster took place on January 28, 1986, changing the Hubble program plans significantly.

7.3.7.4 Questions to the Reader

What could be some immediate impacts to the program?
What could be new priorities for the program?
What should be done during this (undetermined) wait period?

7.3.7.5 Eventual Launch

Launch was delayed during the Space Shuttle return-to-flight redesign and recertification program that followed the Challenger accident. Systems engineers used the interim period to significant program advantage for extensive testing and evaluation to ensure high system reliability and ready feasibility of planned on-orbit servicing maintenance functions.

The telescope was finally transported from the Lockheed site in California to the Kennedy Space Center, Florida, in 1989. It was prepared for launch and carried aloft aboard the Space Transportation System STS-31 mission of the Space Shuttle Discovery on April 24, 1990. This was almost 4 years after the original plan.

The end-to-end test never took place. Historical debate questions the cost estimate of the full test. NASA estimated the cost about $100M alone, based on having to create new facilities. Some have claimed the Air Force could have performed the tests using existing facilities for $10M.

7.3.7.6 Problem with the Primary Mirror

The HST, with anticipated resolution power some 10 times better than any telescopic device on Earth, was on the verge of introducing a whole new dimension of astronomical research and education. However, soon after initial check-out, operations began to show mixed results, a major performance problem was traced to a microscopic flaw in the main mirror that significantly reduced the ability of the telescope to focus properly for demanding (and most valuable) experiments. The focusing defect was found to result from an optical distortion due to an incorrectly polished mirror. This "optical aberration" prevented focusing light into a sharp point. Instead, the light collected was spread over a larger area, creating a fuzzy, blurred image, especially for faintly lighted or weakly radiating objects.

Since the primary mirror could not practicably be returned to earth or physically repaired on orbit, the decision was made to develop and install corrective optics for HST instruments. The idea parallels putting on prescription eyeglasses or contact lenses to correct a person's vision. This approach proved feasible, even if physically and technically challenging, because the system engineers had designed the system specifically for life cycle consideration: on-orbit servicing to upgrade instruments and change out degradable components. Instruments were designed to be installed in standard dresser–drawer fashion for ease of removal and replacement.

On December 2, 1993, the STS-61 crew launched on Space Shuttle Endeavor for an 11 day mission with a record five spacewalks planned. Watched by millions worldwide on live television (Figure 7.10), the astronauts endured long hours of challenging spacewalks to install instruments containing the corrective optics. This was the first opportunity for maintenance, repairs, and upgrades, which included the telescope's solar arrays, gyroscopes, wide-field camera, high-speed photometer, new computer coprocessor, the solar array drive electronics unit, and the Goddard

Figure 7.10 Shuttle mission STS-61 repairs HST optical telescope assembly. (Photo courtesy of NASA.)

high-resolution spectrograph redundancy kit. After 5 weeks of engineering check-out, optical alignment, and instrument calibration, the confirmation of success came as the first images from the space telescope were received on the ground.

7.3.7.7 Learning Principle

Life cycle support planning and execution must be integral from the start of concept refinement and system design.

When originally planned in 1979, the Large Space Telescope program called for return to Earth, refurbishment, and relaunch every 5 years, with on-orbit servicing (OOS) every 2.5 years. Hardware lifetime and reliability requirements were based on that 2.5 year interval between servicing missions. In 1985, contamination and structural loading concerns eliminated the concept of ground return. NASA decided that OOS might be adequate to maintain HST for its 15 year design life. A 3 year cycle of OOS was then adopted. The first HST servicing mission in December 1993 was an enormous success. Additional servicing missions were accomplished in February 1997, December 1999, and March 2002. In 2006, NASA proposed to discontinue servicing Hubble and requested proposals/information to robotically bring it down. Later, this decision was reversed and a fourth service mission was planned in 2009. This complex mission includes replacing all six gyros, adding new batteries, and updating several of the instruments; it is expected to extend the life of Hubble by 5 years to 2015.

The Challenger-imposed 4 year delay provided needed time to look deeper into servicing mission details and to take advantage of other contingencies that might develop. While the primary mirror repair was not one of these, all of the provisions for the others enabled the mirror fix to be accomplished more expeditiously than might have otherwise been the case.

For HST, the most significant correct decision for servicing missions was the agreement that the program would use the physical assets and added time to rigorously work the on-orbit issues. This may seem obvious, but the telescope "owners" are the science community, which understandably, and normally for good reason, does not want anyone to "mess" with their delicate instruments.

Clearly, the time made available by the Challenger delay proved vital and the decision to invest significant additional resources, including SE talent, was both wise and necessary. The effective use of Challenger-imposed downtime allowed mastery of on-orbit astronaut tasks simplify using actual hardware. This decision stands to NASA's credit, as NASA could have easily justified further delay in applying new resources while awaiting the outcome of the Challenger investigation and return-to-flight program plan. The prudent use of the time and additional resources also leveraged improved outcomes for the later program life cycle phase.

HST represents the benchmark for building in system life cycle support (reliability, maintainability, provision for technology upgrade, built-in redundancy, etc.), all with provision for human execution of functions critical to servicing missions. With four successful service missions complete, including one initially not planned for the primary mirror repair, the benefits of design-for-sustainment, or life cycle support, throughout all phases of the program becomes quite evident. Had this not been the case, it is not likely that the unanticipated, unplanned mirror repair could have even been attempted, let alone been totally successful.

7.3.8 Risk Assessment and Management Concept Domain

The SE process must manage risk, both known and unknown, as well as both internal and external. Risk management should specifically identify and access risk factors and their impact, then develop management and mitigation strategies to address them.

7.3.8.1 Early Navigation Solutions

The Global Positioning System (GPS) program evolved as a result of several navigation studies, technology demonstrations, and early operational systems (O'Brien and Griffin, 2008). Sea and air navigation needs during WWII resulted in systems being developed, such as the U.K. GEE system and the U.S. long range navigation (LORAN). These were the first navigational systems to use multiple radio signals to measure the doppler effect (i.e., the difference in the arrival of signals), as a means of determining position. After the Russian Sputnik I launch in 1957, a study by Johns Hopkins University Applied Research Laboratory concluded that a complete set of orbit parameters for a near-earth satellite could be inferred from a single set of

Doppler shift data (single pass from horizon to horizon). Noteworthy, the problem could be inverted by knowing the satellite orbit and inferring the ground location. In the late 1950s, the Navy pursued this approach with the Navy navigation satellite system (known as TRANSIT), which would be used to precisely determine the location of Polaris submarines as an initial condition for missile launch.

The operational configuration of TRANSIT was six satellites in polar orbit at approximately 600 nautical miles. Satellite ephemeris was broadcasted, and the provided navigational solution was two dimensional. As an operational system, TRANSIT pioneered many areas of space technology including stabilization systems and advancing time and frequency standards. Later, in 1972, the Navy continued space navigation research with the TIMATION project. These satellites first used quartz crystal oscillators to provide precise timing. Later, TIMATION launched the first atomic frequency standard clocks.

In the early 1960s, the Air Force then began conducting studies, Project 57 and later Project 621B, on the use of satellites for improving navigation for fast-moving vehicles in three dimensions. One of these studies summarizes the early GPS concept using four orbits and a signal structure using two pseudorandom noise (PNR) sequences. This concept made use of early spread spectrum and would allow use by high-performance aircraft, as well as all the other vehicles requiring navigation information. The faint signal could be detected by users at levels much, much less than that of ambient noise.

The Army was also interested in satellite navigation systems and developed and launched the sequential collation of range (SECOR) system in 1962. The SECOR system continued in use through 1970. With all three services studying and launching satellites for navigation for different needs (e.g., submarines, ships, missiles, people, vehicles, and aircraft), Deputy Secretary of Defense David Packard was concerned about the proliferation of programs being individually pursued by the services. In 1973, he directed that the spaced-based navigation efforts for the three services would become a single joint program. The Air Force was directed to be the lead with multiservice participation. The Joint Program Office (JPO) was created at the Space and Missile System Organization at Los Angeles Air Station and the GPS program planning began.

7.3.8.2 Global Positioning System Acquisition

Soon after the establishment of the JPO, the first major task was to obtain approval and funding for the program. Recall that the services were each doing separate activities. A new program would have to be supported and would impact existing systems. So, the program began on a risky start. The program needed to answer some basic questions. Will a universal system permit a significant reduction in the total DoD cost for positioning and navigation? Will military effectiveness be significantly increased by a universal system? And lastly, what is the best program acquisition strategy and schedule for achieving the desired capability?

Luckily, much of the technological risk had already been mitigated for the GPS program. This included such aspects as space system reliability through the TRANSIT program; the stability of atomic clocks and quartz crystal oscillator through the TIMATION program; the precise ephemeris tracking and algorithms prediction; the spread spectrum signal (PNR) structure from Project 621B; and large-scale integrated circuits in a general industry-wide effort. Reliability of satellites and large-scale integrated circuits had been proven. The resultant GPS program was then a synthesis of the best from each service's programs.

A concept development plan was approved in May 1974 that documented expected GPS performance. Also, it was decided that the program would consist of a three-phase approach:

Phase I—Concept/validation
Phase II—Full-scale engineering development
Phase III—Production

The program estimated a limited IOC could be obtained in 1981 and a full operational capability in 1984.

While there was still no concept of operations, the program manager established a vision of two "performance requirements." The first requirement was to demonstrate "dropping five bombs in the same hole." This reflected the ability to transmit and use GPS data and integrate it into military operations. It also conveyed a demonstration to gain support for the program. The second requirement was to build an affordable receiver, for less than $10,000 a unit.

A number of decisions were made in Phase I that greatly reduced programmatic risk. No additional requirements were added to this phase of the program. From contract award to launch in only 3½ years, there were only two small configuration changes. Next, the acquisition strategy was to issue separate contracts for each segment: space vehicle, control station, and user equipment. And lastly, since much of the technology was mature, fixed price, multiple incentive contracts were used whenever possible. For example, the first four Block I satellites were procured using a fixed-price contract. It did have a 125% ceiling to address problem resolution.

Another interesting aspect to configuration related to the system specifications. GPS Phase I had segment specifications, as expected, for the user equipment, the space vehicles, and the control system. But it had a fourth specification for navigation technology, which focused on validating various technology concepts, especially the space-borne atomic clocks. The navigation technology segment of the GPS provided initial space qualification tests of rubidium and cesium clocks. This segment provided the original test of the GPS signals from space, measurement of radiation effects, longevity effects on solar cells, and initial orbital calculations. As part of a six nation cooperative experiment, precise time synchronization of remote worldwide ground clocks was obtained using the first two GPS technology satellites (NTS-1 and NTS-2) during May through September 1978 (see Figure 7.11).

Figure 7.11 **Navigation technology satellite. (From U.S. Navy, *TIMATION and GPS History*, http://NCSTwww.NRL.navy.mil/NCSTOrigin/TIMATION.html (accessed October 4, 2007.)**

The GPS JPO decided to retain core SE and system integration responsibility. Thus, a major cornerstone of the program was the interface control. The integration role required contact with many government and industry entities. A plethora of technical expertise, test organizations, users/operators, service acquisition offices, etc. required working interfaces and integration. Managing this complex set of stakeholders successfully was a major programmatic risk.

7.3.8.3 Questions to the Reader

What risks were present throughout the described Phase I GPS program?
Is there risk to a program with so many stakeholders and external organizations?

Could the commercial aspects of the user equipment have been predicted or planned?

What could be technical risks in Phase II, or in later phases?

7.3.8.4 Learning Principle

Disciplined and appropriate risk management must be applied throughout the life cycle.

The GPS program is rich with many examples of successful risk management. However, funding was reduced to essentially zero for the Phase III (Block II) production satellites. This would greatly affect how many satellites would be in the final GPS constellation, subsequently how many satellites a user would be able to "see." Manufacturing sources for the atomic clocks became an issue, as did the ability to launch GPS space vehicles, after the Space Shuttle Challenger disaster in January 1986.

The GPS program was structured to address risk in several ways throughout the multiphase program. Where key risks were known up front, the contractor and/or government utilized a classic risk management approach to identify and analyze risk, and then developed and tracked mitigation actions. Identified technical risks were addressed at weekly chief engineer meetings, and often tracked by technical performance measures.

The program office, serving in the clear role of integrator, would issue a RFP to several bidders for developing concepts and/or preliminary designs. Then, one contractor would be selected to continue development. This approach not only provided innovative solutions through competition but also helped in defining a lower risk and more clearly defined development program. It also allowed for, if appropriate, fixed-price contracts to constraint program cost.

Risk management is a disciplined approach to dealing with uncertainty that is present throughout the entire systems life cycle. It addresses risk planning, risk assessment, risk handling and mitigation strategies; and risk monitoring approaches. Most recently, the concept of opportunity is often discussed with risk analysis. The objective is to achieve a proper balance between risk and opportunity.

7.3.9 Systems and Program Management Concept Domain

Systems and program management spans the variety of activities within the ninth, and last, concept domain of the F–S framework. It provides a balancing and control function that anchors SE process execution. It also ensures a proper connection to other business and financial management processes that combine to yield a successful program. Rather than a single vignette for systems and program management, what follows is a short synopsis of selected learning principles from several programs. Collectively, these learning principles cover a broad array of subjects that illustrate the expansive nature of systems and program management. More specific details can be found in the individual case studies (Garland, 2008).

7.3.9.1 Use of Independent Review Teams Is Beneficial

The Air Force C-5 acquisition program office employed independent review teams (IRTs) to assemble national experts to examine the program and provide recommendations to the government (Griffin, 2005). These problem-solving teams were convened to garner the best advice in particular technical areas: such as structure design or service life.

The Air Force used IRTs to augment their staff, assess difficult problems, and recommend solutions to problems. The program office had the power, ability, priority, and senior leadership support to quickly rally the best minds in the nation. The SE process developed the agenda, established the teams' charter, collected data for review, assisted in developing the trade study information, and aided in constructing alternative solutions. The IRTs presented the matrix of solutions for final selection and approval.

7.3.9.2 Communications and Systems Management Are Critical

The F-111 suffered from poor communications between the Air Force and Navy technical staffs, and from overmanagement by the Secretary of Defense and the Director, Defense Research and Engineering, and it came under intense congressional scrutiny, which restricted the program manager from applying sound SE principles (Richey, 2005).

7.3.9.3 Systems Engineering Team Integrity Must Be Maintained across the Functional Hierarchy of the Organization

The B-2 contract stipulated a work breakdown structure (WBS) for the entire program content (Griffin and Kinnu, 2005). The company organized the design and development effort into multiple teams, each responsible to implement the WBS for subsystems of the air vehicle. These WBS task teams were assigned complete work packages, for example, the forward center wing. The SE WBS task team efforts were organized similarly, but with separate systems organizations, each reporting to the Northrop chief engineer. The functional organizations assigned members to the task teams to assure accommodation of their program needs. A vital distinction from many of today's IPTs was retaining the WBS task team membership throughout the functional organizations' various management levels. This facilitated communication, integration, interfaces, and integrated the functional leadership of each technical and management discipline into the decision process. The program management top-level structure was organized into a strong project office with centralized decision authority at the top.

7.3.9.4 Programs Must Strive to Staff Key Positions with Domain Experts

From program management to SE, to design, to the manufacturing and operations teams, the people on the GPS program were well versed in their disciplines, and all possessed a systems view of the program (O'Brien and Griffin, 2008). While communications, working relationships, and organization were important, it was the ability of the whole team at all levels to understand the implications of their work on the system that was vital. Their knowledge-based approach for decision making had the effect of shortening the decision cycle, because both the information was understood as well as the base and alternative solutions were accurately presented.

7.3.9.5 Government Must Ensure the Contractor Is Able to "Walk the Talk" When It Comes to Production

In reviewing the abilities of the contractor to execute the contract for the A-10, the Air Force failed to identify a number of issues that might well have doomed the program to failure. Both before and after the awarding of the contract, the company was in trouble. The government's pre-award survey of Fairchild Hiller examined their capacity, capability, and financial condition but failed to recognize some of the risk elements and concerns that would be noted some 3 years later. Fairchild had failed to adequately invest in equipment and its workforce, and its management and organization had deficiencies. The Air Force made a number of recommendations that were followed, and the company was able to produce the aircraft; however, it ultimately put the company in a position from which it could not recover (Jacques and Strouble, 2008).

7.3.9.6 Investing in Long-Term Systems Engineering and Program Management Staff Supports Program Success

The number and types of personnel assigned to the Peacekeeper program; high levels of knowledge, skill, ability, and experience of the personnel, and lengthy tenure assignments all contributed tremendously to the success of the program. The engineering and advanced degrees held by most of the leadership were critical in enabling them to make informed decisions and manage their programs. The engineers were allowed sufficient time to learn missile systems, gain experience in all aspects and subsystems, and learn and practice the Ballistic Missile Office (BMO) systems engineering process. The BMO staff had the manning and expertise to carefully monitor subcontractors and actually participate in detailed reviews, allowing them to make all critical decisions for the best outcome of the Air Force. Adequate manning, experiential learning, supportive work environment, hands-on experience, job rotation, mentoring, hand-picked staff, and staff education level were key factors in the success of the Peacekeeper program (Stockman and Fornell, 2008).

7.3.10 Summary

Attributed to Sir Francis Bacon, sixteenth-century philosopher, scientist, lawyer, and statesman, "History makes people wise." Case studies, perhaps better than any other tool, can provide the student an opportunity to learn from those that have gone before us. For the educator, case studies provide the opportunity to step away from a lecture mode and to plunge the student into participative dialogue. For the practitioner, case studies can help prevent the endless, needless repetition of mistakes. It is the authors' sincere hope that this chapter increases, in some small way, the collective SE-related wisdom of students, educators, and practitioners.

Authors

Charles M. Garland, is senior consultant at the Air Force Center for Systems Engineering, Wright-Patterson Air Force Base, Ohio. He is the manager for case study development and was technical editor for the Global Hawk, Peacekeeper Intercontinental Ballistic Missile, International Space Station, and T-6 Texan II case studies. Charles has over 30 years of engineering experience in the Department of Defense acquisition environment on such programs as the F-16 fighter, B-52 and B-1 bombers, and Global Hawk unmanned aircraft system. Garland has a master's degree in systems engineering from Wright State University and is a registered professional engineer in the State of Ohio.

John Colombi, an assistant professor of systems engineering at the Air Force Institute of Technology (AFIT). His research interests include systems and enterprise architecture, complex adaptive systems, and human systems integration. Before joining AFIT, Dr. Colombi served 21 years in the active duty U.S. Air Force where he led various C4ISR systems integration and systems engineering activities. He developed information systems security at the National Security Agency and researched communications networking at the Air Force Research Laboratory. Dr. Colombi managed and edited the first five case studies for the Air Force Center for Systems Engineering; these included the B-2, C-5, Theater Battle Management Core Systems (TBMCS), F-111, and the Hubble Space Telescope.

References

Friedman, G. and A. Sage. 2004. Case studies of systems engineering and management in systems acquisition. *Journal of Systems Engineering* 7(1): 84–97.
Garland, C. 2008. *Systems Engineering Case Studies: Synopsis of the Learning Principles.* Wright-Patterson AFB, OH: Air Force Center for Systems Engineering, http://www.afit.edu/cse/cases.cfm (accessed December 9, 2009).

Garvin, D.A. 2003. Making the case: Professional education for the world of practice. *Harvard Magazine* 106, (1): 56.

Griffin, J.M. 2005. *C-5A Galaxy Systems Engineering Case Study.* Wright-Patterson AFB, OH: Air Force Center for Systems Engineering, http://www.afit.edu/cse/ (accessed December 9, 2009).

Griffin, J.M. and J.E. Kinnu. 2005. *B-2 Systems Engineering Case Study.* Wright-Patterson AFB, OH: Air Force Center for Systems Engineering, http://www.afit.edu/cse/ (accessed December 9, 2009).

Jacques, D.R. and D.D. Strouble. 2008. *A-10 Thunderbolt II (Warthog) Systems Engineering Case Study.* Wright-Patterson AFB, OH: Air Force Center for Systems Engineering, http://www.afit.edu/cse/ (accessed December 9, 2009).

John F. Kennedy Presidential Library & Museum. 2009. *Biographies & Profiles: Robert McNamara,* http://www.jfklibrary.org/ (accessed December 9, 2009).

Joint Chiefs of Staff. 2009. *Joint Publication 3-09.3: Close Air Support (CAS).* Washington DC: Department of Defense.

Launius, R. and B.J. Dvorscak. 2001. *The C-5 Galaxy History Crushing Setbacks, Decisive Achievements.* Paducah, KY: Turner Publishing Company.

Mattice, J.J. 2005. *Hubble Space Telescope Systems Engineering Case Study.* Wright-Patterson AFB, OH: Air Force Center for Systems Engineering, http://www.afit.edu/cse/ (accessed December 9, 2009).

NASA. 2007. *NASA Systems Engineering Handbook.* NASA/SP-2007–6105 Rev 1.

O'Brien, P.J. and J.M. Griffin. 2008. *Global Positioning System Systems Engineering Case Study.* Wright-Patterson AFB, OH: Air Force Center for Systems Engineering, http://www.afit.edu/cse/ (accessed December 9, 2009).

Richey, G.K. 2005. *F-111 Systems Engineering Case Study.* Wright-Patterson AFB, OH: Air Force Center for Systems Engineering, http://www.afit.edu/cse/ (accessed December 9, 2009).

Stockman, B. and G.E. Fornell. 2008. *Peacekeeper Intercontinental Ballistic Missile Systems Engineering Case Study.* Wright-Patterson AFB, OH: AF Center for Systems Engineering, http://www.afit.edu/cse (accessed December 9, 2009).

United States Air Force (USAF). 2009. *Systems Engineering Case Studies.* http://www.afit.edu/cse (accessed December 9, 2009).

United States Navy (U.S. Navy). 2007. *TIMATION and GPS History.* http://NCSTwww.NRL.navy.mil/NCSTOrigin/TIMATION.html (accessed October 4, 2007).

Chapter 8

Integrated Reliability Management System for Faster Time-to-Market Electronic Products

Tongdan Jin and Madhu Kilari

Contents

This chapter proposes an integrated reliability management system for faster time-to-market electronics systems in the context of distributed manufacturing paradigm. A major benefit of distributed manufacturing is cost reduction, which makes products more competitive. However, the distributed manufacturing environment introduces many risks into product quality and reliability. Our objective is to address the reliability issue across the product life cycle that encompasses equipment design, manufacturing, integration, and field usage. The idea is to bridge the reliability information gap between the manufacturer and the customers through an effective failure tracking and corrective action (CA) system. Engineers can use the information system to carry out reliability analysis, implement CA, and track the reliability growth rate. The system consists of four core modules: (1) a stochastic-based reliability prediction model incorporating both component and noncomponent failures; (2) real-time failure mode (FM) run charts; (3) CA effective functions; and (4) reliability monitoring metrics. The proposed reliability management system will be demonstrated on the automatic test equipment (ATE). Our research shows that reliability growth can be effectively achieved using the entire product life cycle approach. This is particularly important for new products designed for faster time-to-market.

8.1 Introduction

Reliability growth can be achieved at the system level if extended in-house testing is allowed. In a distributed manufacturing environment, product design, manufacturing, and integration are carried out at different spatial and temporal scales, leading to many uncertainties in the final product (Chen and Sackett, 2007). The distributed manufacturing paradigm is built upon the pipeline model. That is, the market pushes the design, the design pushes the manufacturing, and the manufacturing further pushes the final shipment. Hence, the new product can be released to the market in the shortest possible time. Throughout the chapter, product and system will be used interchangeably.

In semiconductor, telecommunication, and other technology-driven industries, the requirement of high system reliability render traditional reliability management methods obsolete due to the shift of the manufacturing paradigm. Take the ATE systems as examples. This type of multimillion-dollar equipment is widely used to test wafers and electronic devices in semiconductor industry. Often the design of this complex equipment is undertaken by engineers in the United States while the hardware assembly, software development, and final integration are accomplished in low-cost regions outside the United States. A major benefit of the distributed manufacturing chain is the cost reduction, yet it potentially introduces many uncertainties into product quality and reliability.

Figure 8.1 depicts a generic ATE system configured with eight instrument modules for testing two devices at the same time. There are five types of instrument modules in this system: high-speed digital (HSD), analog, direct current (DC), radio frequency (RF), and support modules. Each instrument module is actually built upon a printed circuit board (PCB) that contains thousands of electronic components to form various circuits. Redundant designs are usually not implemented due to the area limitations on the PCB. The device interface board (DIB) provides electrical connections between the instrument modules and the device under test (DUT). The system configuration may vary depending on the type of DUT being tested. For example, if the system is used to test logic devices requiring pure digital singles, more HSD are installed in the system.

When the system is developed and manufactured in the distributed supply chain, system-level reliability evaluation cannot be effectively performed until the system is assembled prior to the shipment. Due to the diversity of customer usages, many latent failures or design weaknesses will not show up until the system has operated in fields for several months or even a year. Hence, the system manufacturer

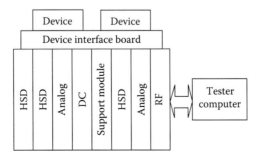

Figure 8.1 A generic ATE system for dual sites testing.

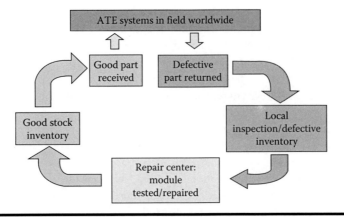

Figure 8.2 Repairable PCB module transition loop.

often continues to improve the system's reliability by collecting failure data, analyzing failure modes (FM), and implementing corrective action (CA) after the field installation. Figure 8.2 depicts the flow chart of the PCB repair activities.

Complex systems like ATE are often designed in modules to facilitate repair and maintenance activities. Upon failure, the defective module is swapped with a spare module, and the system is restored to production immediately. Since each PCB module is very expensive typically costing between $5,000 and $10,000, the defective module is often sent back to the repair center for root-cause analysis. Once the problem is fixed, it is sent back to the good stockroom for replacement in case of another failure in the future. The annual flow rate of defective modules is high when the reliability of new equipment is lower. Assuming 1500 systems are installed worldwide, and the system mean time between failure (MTBF) is 2000 h, the annual field returns will reach 6570 modules. This implies a large amount of repair and warranty cost to the ATE manufacturer.

The traditional reliability management methods become ineffective in the distributed manufacturing environment. Firstly, the uncertainties in product reliability are amplified due to the involvement of subcontractors who may have less stringent reliability and quality specifications. Secondly, system failure information is prone to delay either due to geographical separation or communication barriers between the manufacturer and its global customers. For instance, it usually takes couple of days or even two weeks for a defective module to return for repair. Proactive measures or containment activities cannot be initiated immediately upon the occurrence of major FMs. Thirdly, information flow between the manufacturers and their customers is often poor. Customers are not able to directly access the system reliability information, which discourages data sharing between the manufacturer and the customers. Therefore, increasing product reliability becomes the sole responsibility of the manufacturer, while customers are only interested in the warranty services. Finally, the manufacturer often has difficulties in prioritizing the limited resources to improve the system reliability when multiple failures occur at the same time.

Our research aims to propose an integrated reliability management system that explicitly considers design uncertainties, process variations, and diversities in field usage. The research directly explores the theoretical domain to advance reliability prediction techniques and further the development of effective reliability tracking methods by incorporating hardware, software, design, manufacturing, and process-related issues. The ultimate goal is to establish a reliability growth planning (RGP) framework that is able to achieve continuous reliability improvement across the product lifetime when products are manufactured in distributed supply chains.

The rest of the chapter is organized as follows. In Section 8.2, the challenges to traditional reliability management will be discussed, and the concept of the integrated reliability management system is introduced. In Section 8.3 a stochastic-based reliability prediction model is proposed. Section 8.4 focuses on the derivation of real-time FM run charts. In Section 8.5, CA effectiveness function is proposed to link the monetary cost with the expected failure reduction rate. In Section 8.6, a failure-in-time (FIT) based reliability growth monitoring system is discussed. Section 8.7 concludes the chapter with some discussions on future research work.

8.2 Challenges in Reliability Growth Planning

Design for reliability in electronic products became possible with the introduction of reliability prediction tools in 1960s. Two widely used tools are Mil-HDBK-217 (US MIL-HDBK-217, 1965) and Bellcore/Telcordia TR-332 (Telcordia Technologies, 2001). Based on these two standards, various commercial software applications were developed to facilitate the estimation of product reliability. Including the Mil-HDBK and Bellcore standards, most MTBF prediction methods were proposed based on the component failure rates and the bill-of-materials (BOM) of the product.

With regard to electronic system design, the existing reliability prediction models face three major challenges: (1) the increase of noncomponent failures; (2) compressed product design cycles; and (3) diverse product usage profiles. Noncomponent failures occur due to design errors, software bugs, manufacturing defects, and improper field usages. The rapid progress of technology makes the electronic products obsolete every 3–5 years, which pushes the manufacturers to release the product in the shortest time window to gain market share. The system usage profile at customer sites varies depending on the devices being tested. Although the system is designed with certain protection mechanisms or guard bands, the unintentional programming or human error still breaks the protection barrier and causes the damage to the system during field operation.

8.2.1 Noncomponent Failures

Accurate reliability prediction such as MTBF is always desirable and anticipated before the new product is ramped up for volume manufacturing. System MTBF estimated from component failure rates is quite optimistic in general. This is particularly true for a new product design when it is developed for a faster time-to-market requirement. In fact, a PCB module could fail because of issues related to defective components or noncomponents. Component failures primarily belong to defective hardware devices. Noncomponent failures include design errors, manufacturing issues (e.g., poor solder joints), software bugs, and improper customer usage. For some new products, noncomponent failures could dominate the failure rates in the early product life cycle. Therefore, the new product often exhibits a lower MTBF than what is anticipated based on only component failure rates.

Table 8.1, originally from Jin et al. (2006), summarizes the failure data of four types of PCB product lines. Data are collected from dozens of field systems during 1 year period since the shipment of the first system. Failures are classified by root-cause categories: component, design, manufacturing, process, software, and no-fault-found (NFF). An interesting observation is that component failures, though the largest, contribute less than 42% to total failures. For PCB-B, component failures account for only 26% of field failures. In fact, the largest failure arises from NFF, which contributes 27% to failures. NFF is a kind of failure occurring at the customer site or observed by customers, but the FM cannot be duplicated by the manufacturer.

Figure 8.3 presents the MTBF run chart for PCB-A after it was released to the market (Jin et al., 2006). The product reliability exhibits a growth trend due to continuous CA activities. The reliability target for this product is 40,000 h MTBF.

Table 8.1 Failures Breakdown by Root-Cause Categories

Failure Category	PCB Module Type			
	A	B	C	D
Component (%)	42	26	34	28
Design (%)	9	18	21	33
Manufacturing (%)	12	10	7	4
Process (%)	7	13	16	13
Software (%)	2	6	2	7
NFF (%)	28	27	20	15
Total (%)	100	100	100	100

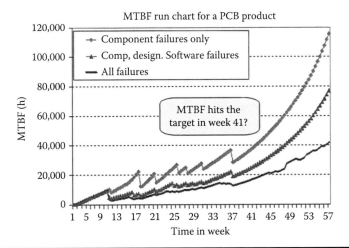

Figure 8.3 **Actual product MTBF vs. predictions.**

Based on the traditional prediction model (i.e., component failures only), the reliability target could be reached by week 41. When the prediction model includes both components and software failures, the product will reach the target in week 48, that is, delayed by 7 weeks. It actually took 57 weeks to reach the target MTBF after we consider all types of failures including cold solder joints, design issues, and process-related problems. The chart clearly shows if we ignore the noncomponent failures, the resulting reliability prediction would be too optimistic compared to the actual product MTBF.

Table 8.1 and Figure 8.3 indicate that accurate reliability prediction is difficult for a complex hardware–software system, yet it is essential for the product manufacturer to allocate necessary resources to improve reliability and reach the reliability goals. On the other hand, optimistic prediction could potentially mislead the management to withdraw engineering resources early from reliability monitoring and CA implementation. High reliability is also important to obtain high levels of customer satisfaction. Lower system reliability implies more system downtimes at the production floor, which could mean a loss of hundreds of thousands of dollars to the customer due to idle labor, unfinished testing devices, and equipment depreciation.

8.2.2 Compressed Design Cycles

If a system consists of different types of modules each having different development times, it is extremely difficult, if not impossible, to assemble all types of modules and conduct a system level test. In the semiconductor industry, new generation of devices are often released to the market every 18–24 months according to Moore's

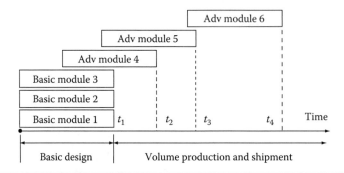

Figure 8.4 Unsynchronized module development times.

law. The introduction of a new device often requires the testing equipment to be upgraded to meet the new testing performance. This is often done by replacing old PCB modules with advanced ones. To meet the customer requirement, the equipment manufacturer often ships a new ATE system after the basic modules are incorporated, while advanced modules are still in development or pilot line phases. Figure 8.4 graphically describes the unsynchronized module development strategy. When systems are designed and marketed in such a compressed design schedule, it is almost impossible for the system manufacturer to implement in-house reliability growth testing (RGT) at the system level.

8.2.3 Diversities in Customer Usages

Systems like ATE are often used by hundreds of customers around the world, and each customer site may install dozens of systems. Customer usage contributes to many unexpected system failures. For example, inappropriate device programming by the engineers at the customer often is a major reason for excessive ATE system failures. In reality, the customer usage profiles are very diverse because electronic devices vary in logic, memory, analog, and RF chips. As such, field systems experience different levels of operational stresses.

Some failures or design weakness will not show up until the system has been used in the field for months or a year. Such failures are called dormant failures or latent failures. Latent failures are often induced by uncertainties in design, manufacturing, and operations. Various factors, such as customer usage, design weakness, software bugs, and electrical statistic discharge, may trigger latent failures. Jin et al. (2010) found that latent failures could significantly jeopardize system reliability due to improper customer usage and immature designs.

In general, many reliability prediction models tend to be optimistic, which fails to guide the system manufacturer in allocating CA resources. When a complex system consists of multiple types of modules, and each type of module has a different development time, it is even impossible to assemble the modules and conduct a system-level test. Obviously, reliability prediction and management for such types of complex systems motivate the development of new reliability growth approaches.

8.2.4 Current Researches

System reliability growth can be achieved through RGT if the design is simple or extended life testing is feasible. The idea of RGT dates back to Duane (1964) in 1960s when he was in charge of monitoring the lifetime of aircrafts. Crow (1974) found that reliability growth curve can be modeled by a nonhomogenous Poisson process (NHPP). His findings eventually were summarized and published as the popular Crow/AMSAA model. Since then, significant research activities were carried out in RGT (Crow, 1984, 2004; Clark 1999; Wang and Coit, 2005). Recently, Krasich et al. (2004) and Krasich (2006) described an accelerated RGT method during product development stage. These researches usually focus on the reliability growth issue for new products or systems during the design and development phase assuming extended in-house testing time is available.

RGT becomes ineffective if the system becomes complicated or the new design is driven by faster time-to-market shipment. In that case, RGP can be implemented to achieve the reliability goal across the entire product lifetime. Compared to RGT, the concept of the RGP is broader in the sense that the latter is able to drive the reliability growth across product design, manufacturing, assembly, and field use. In RGP, the manufacturer collects field data, analyzes failure mechanisms, and implements necessary CA in a timely manner. Smith (2004) described a RGP methodology for planning and estimating the cost of a reliability improvement program through CA based on field data. Ellner and Hall (2006) proposed a parsimonious approximation method to develop reliability growth plans based on CA against known FMs. Through RGP, the system failure intensity can be reduced and the reliability goal will be achieved within a predetermined time.

8.2.5 Proposed Methods

This chapter proposes an integrated reliability management system to improve field system reliability when a new system is designed and developed in a compressed schedule. We address the reliability management issue from the systems point view. The idea is to bridge the reliability information gap between the product manufacturer and the customers through a failure information sharing system. The research will directly explore the theoretical domain to advance reliability prediction techniques and introduce a real-time FM run chart by incorporating reliability growth and demand uncertainties.

Figure 8.5 graphically presents the idea of the new reliability management system. The integrated reliability management system creates an information bridge among the product manufacturer, the repair center, and the field customers. The

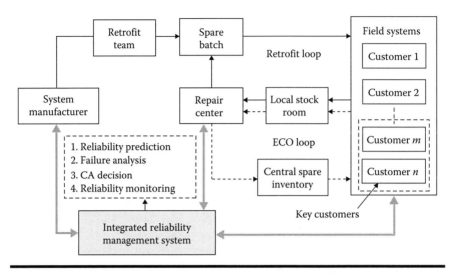

Figure 8.5   Integrated reliability management system.

manufacturer is able to monitor the product reliability and prioritize the CA activities based on the information provided by the reliability management system. In practice, two types of CA are often used: retrofit and engineering-change-order (ECO). Retrofit is often executed by a special team that proactively uses spare modules to replace field modules, which may fail due to a known FM. ECO is a countermeasure generally implemented in the repair center to upgrade defective modules by replacing certain components prior to failures.

As shown in Figure 8.5, four reliability management tools are proposed as integral parts of the reliability management system: (1) stochastic-based reliability prediction models incorporating both component and noncomponent failures; (2) real-time FM run charts considering the change of field product population; (3) CA-effective functions; and (4) reliability growth monitoring metrics. In Sections 8.3 through 8.6, these tools will be systematically described and demonstrated on the ATE.

8.3 Stochastic-Based Reliability Prediction

A good failure data tracking system is essential for achieving accurate reliability predictions. To create and maintain an accurate reliability information database, dedicated personnel are needed to track and record the failure data reported by the repair center. For example, Honeywell (Gullo, 1999; Johnson and Gullo, 2000) allocated specific resources to maintain a consistent database, which systematically tracks and records product failures due to design, manufacturing, and field services. As the manufacturing industry continues to move offshore driven by the competitive manufacturing cost outside the United States, creating such a type of failure data tracking system becomes increasingly important. The database will serve as a basis on which the manufacturer can engage in an information-rich conversation with the component suppliers in case of any reliability and quality issues arising from field systems.

The database system also facilitates the reliability prediction of new product designs. For components used in the new PCB module, some have known failure rates as they can be derived from historical field data. Others may not have explicit failure rates because they are new and used for the first time in the design. In that case, the nominal component failure rate can be used to forecast the actual failure rate by incorporating the actual operating condition. Unlike component failures, failure rates of noncomponents are more project-specific. Triangle distributions will be adopted to model the uncertainty in noncomponent failure rates. In Sections 8.3.1 through 8.3.3, the models proposed by Jin et al. (2006) will be reviewed.

8.3.1 Component Failure Rate

Let $\hat{\lambda}_c$ be the failure rate estimate of a particular component. The system shipment information enables us to estimate component cumulative operating hours in the field. The number of defects for that particular component is often available from failure database. Then the component failure rate can be appropriately estimated by

$$\hat{\lambda}_c = \frac{\text{total component failures}}{\text{cumulative operating hours}} \qquad (8.1)$$

For instance, a type of 5 V relay is used in a system design. Each system employs 10 relays and there are 200 systems in the field. During 1 year period, five field failures were caused by defective relays. Assuming the system operates continuously throughout a year (8760 h), then the failure rate for the relay can be estimated by

$$\hat{\lambda}_c = \frac{5}{8760 \times 10 \times 200} = 2.85 \times 10^{-7} \text{ failures/h} \qquad (8.2)$$

If this type of relays will be used in a new product design, then the expected failure rate for the relays can be estimated as 2.85×10^{-7} failures/h if the operating condition is similar to the predecessor.

For a new component used in the system design, the actual failure data usually is not available. Component suppliers often provide the nominal failure rate denoted as $\hat{\lambda}_0$, which is estimated in the ambient temperature of $T = 40°C$ and electrical derating $p = 50\%$. If the operating temperature and derating are higher than the nominal condition, the actual component failure rate, denoted as $\hat{\lambda}_c$, could be appropriately extrapolated as following:

$$\hat{\lambda}_c = \hat{\lambda}_0 \pi_T \pi_E \tag{8.3}$$

where

$$E[\pi_T] = \int_{T_l}^{T_u} e^{\frac{E_a}{k}\left(\frac{1}{T_0} - \frac{1}{x}\right)} f_T(x)dx \tag{8.4}$$

$$E[\pi_E] = \int_{p_l}^{p_u} e^{\beta(x - p_0)} f_p(x)dx \tag{8.5}$$

Notice that π_T and π_E represent the temperature and electrical derating coefficients, respectively. T_u, p_u, T_l, and p_l represent the upper and low limits for temperature and electrical derating. $f_T(x)$ and $f_p(x)$ are probability distribution functions for temperature and derating. If the system is in early design and development phase, prototypes are not available for estimating the temperature and derating distributions. Then noninformative statistics such as the uniform distribution can be used to model the distributions of T and p. Given Equations 8.3 through 8.5, the mean of $\hat{\lambda}_c$ can be estimated as

$$E[\hat{\lambda}_c] = \hat{\lambda}_0 E[\pi_T] E[\pi_E] \tag{8.6}$$

In summary, when components are new and no field failures are available, the nominal failure rate $\hat{\lambda}_0$ can be appropriately used to extrapolate the actual failure rate by incorporating electrical and temperature stresses. Quite often, the actual operating environment is different from the nominal condition. Then Equations 8.3 or 8.6 can be used to predict the actual component failure rate.

8.3.2 Noncomponent Failure Rate

The growth of the system complexity continuously shifts the reliability attention from the component to the noncomponent domain. Unlike component failures, failure rates of noncomponents are more project-specific. For example, design problems are closely related to the experience of engineers and the complexity of the product. On the other hand, solder joint defects are highly correlated with number of solder joints on the PCB and the reflow temperature profile. The larger the number of solder joints, the more likely the defective solders could occur. Similar to the component failure rate, the noncomponent failure rate γ is defined as

$$\hat{\gamma} = \frac{\text{noncomponent failures}}{\text{cumulative system hours}} \tag{8.7}$$

Using the same example in previous section, if three PCB modules returned from the field and their root causes for failure are identified as cold solder joints, belonging to the manufacturing category, then the manufacturing failure rate, denoted by $\hat{\gamma}_m$, can be estimated as

$$\hat{\gamma}_m = \frac{3}{8760 \times 200} = 1.712 \times 10^{-6} \text{ failures/h} \tag{8.8}$$

Literature on the modeling of noncomponent failure rates is relatively scarce. In reality, available data for noncomponent failures are very limited and they are often project-specific. These factors may explain why many existing reliability prediction models are not able to incorporate noncomponent failures effectively.

The triangular distribution can be used for a subjective modeling of a population for which there is limited failure data or no data available (http://www. brightonwebs.co.uk/distributions/triangular.asp). Hence, it is a good statistical tool to estimate the failure rate distribution of noncomponents. The triangular distribution is represented by three parameters: a, the smallest possible value; b, the largest possible value; and h, the mode. Figure 8.6 shows two probability functions of the unknown failure rate γ denoted as $g_1(\gamma)$ and $g_2(\gamma)$. The analytical expression is given by

$$g(\gamma) = \begin{cases} \dfrac{2(\gamma - a)}{(h - a)(b - a)} & a \leq \gamma \leq h \\[2mm] \dfrac{2(\gamma - b)}{(h - b)(b - a)} & h < \gamma \leq b \end{cases} \tag{8.9}$$

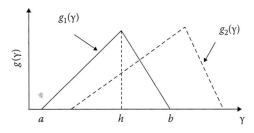

Figure 8.6 Triangular distribution of noncomponent failure rate γ.

where

$$h = 3\bar{\gamma} - b - a \tag{8.10}$$

Notice that $\bar{\gamma}$ is the sample mean of the data set or the average failure rate. Then the mean of the noncomponent failure rate γ can be obtained as

$$E[\hat{\gamma}] = \int_a^b x g(x) dx = \frac{a^3(h-b) + b^3(a-h) + h^3(b-a)}{3(b-a)(h-a)(h-b)} \tag{8.11}$$

For example, assuming failure rates of the manufacturing defects in three existing product lines are: 1.1×10^{-6}, 1.7×10^{-6}, and 2.5×10^{-6} (faults/h). We can immediately obtain $\bar{\gamma} = 1.77 \times 10^{-6}$, $a = 1.1 \times 10^{-6}$, $b = 2.5 \times 10^{-6}$, and $h = 1.7 \times 10^{-6}$. If the new design has similar complexity as its three predecessors, the failure rate distribution for the manufacturing defect can be specified by those estimated a, b, and h.

If the complexity of the new design doubles, failure opportunities for noncomponents will likely double too. For instance, solder joint failures often depend on the number of solder joints in the PCB. When the amount of sold joints on the new PCB module increases by 100%, the failure opportunities for cold solder issues will likely double as well. Therefore, a, b, and h should be appropriately adjusted to accommodate the complexity of the new design.

8.3.3 System Failure Rate

Assuming the system has k types of components, and each component type again has n_i units for ($i = 1, 2, \ldots, k$) in the system. Then the system failure rate, denoted as $\hat{\lambda}_s$, is equal to the summation of failure rates of all components and noncomponents. That is,

$$\hat{\lambda}_s = \sum_{i=1}^{k} n_i \hat{\lambda}_i + \sum_{j=1}^{5} \hat{\gamma}_j \tag{8.12}$$

In (8.12), the first summation represents the cumulative component failure rates. The second term represents the cumulative failure rate for all noncomponent issues. Notice that $j=1$ for design, 2 for manufacturing, 3 for software, 4 for process, and 5 for NFF. $k+5$ is usually a large value as $k \gg 1$. Based on the central limit theorem, $\hat{\lambda}_s$ tends to be normally distributed with the mean

$$\mu_s = E[\hat{\lambda}_s] = \sum_{i=1}^{k} n_i E[\hat{\lambda}_i] + \sum_{j=1}^{5} E[\hat{\gamma}_j] \tag{8.13}$$

8.3.4 *Failure Rate Curves*

Component failure rates often exhibit the so-called bathtub curve as shown in Figure 8.7. The curve consists of three periods: an infant mortality period with a decreasing failure rate, followed by a constant and lower failure rate period (also known as "useful life"). The curve ends up with a wear-out period that exhibits an increasing failure trend. The two-parameter Weibull distributions can be used to model the component lifetime in these different phases. When the Weibull shape parameter is less than 1, it can mimic the decreasing failure rate curve. If it is greater than 1, the Weibull distribution is approximating the wear-out period. When the shape parameter is equal to 1, it becomes an exponential function representing the useful lifetime of the product.

Exponential distributions are widely used to model the lifetime of electronic products such as PCB modules. One reason for its popularity is its mathematical convenience. Another reason is many electronic products often go through the environmental screening process to weed our infant mortality before they are shipped to customers. Since the screening process is so widely practiced, it almost becomes an industry standard to remove infant mortality. Hence the failure rate during the field operation can be treated as a constant number or near a constant rate as given by Equation 8.13.

Unlike component failures, noncomponent failures usually exhibit a decreasing failure rate as shown in Figure 8.7. For example, manufacturing issues such as cold solder joint usually are severe in the early production phase. With tweaking of the reflow temperature, products will have less and less solder joint failures as time goes

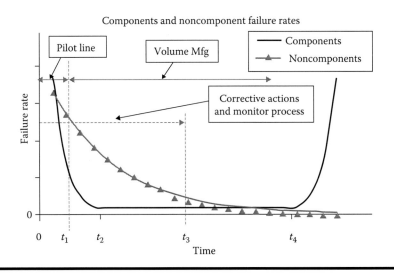

Figure 8.7   Failure rate bathtub curve.

on. The decreasing pattern is also applicable to other noncomponent failures such as design, software bugs, and process-related issues.

8.4 Real-Time Failure Mode Run Chart

Reliability growth for in-service systems can be achieved by implementing CA against major FM. In practice, two types of CA are often used: retrofit and ECO (see Figure 8.5). Retrofit is a proactive measure to replace in-service PCB modules that will fail (but not failed yet) due to known FM. Retrofit is always executed at customer site by swapping out all problem modules with the replacement of good spares. Retrofit can also be applied to software upgrade by issuing a new version to customers.

ECO can be treated as a countermeasure generally implemented in the repair center to fix and upgrade defective modules upon field returns. For example, a PCB is returned from a field system because of a defective capacitor. Upon fixing the capacitor, the technician can proactively replace old relays with new versions per the design upgrade even though these relays are still functional. Hence, ECO is also called self-retrofit in the sense that the CA is applied only if the module is failed and returned for repair. Meanwhile the root-cause and FMs can be systematically recorded in the reliability database system maintained by the repair center. The information could be used for component and noncomponent failure rate estimations or as a reference for component choice in a new system design.

8.4.1 Deficiency of Pareto Chart

CA decisions are often made base on the FM Pareto chart. Figure 8.8 shows a Pareto chart containing six FM with 14 field failures for a type of PCB modules in

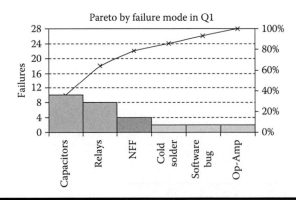

Figure 8.8 Failure mode Pareto chart in the first quarter.

the first quarter (i.e., 13 weeks per quarter). FM are ranked from the left to the right based on the quantity of failure. The chart shows that bad capacitors and relays are two largest FM with total of 18 failures. Based on the chart, decision can be made to prioritize CA against bad capacitors and relays because both represent 65% of the total failures during the 13 week period.

FM charts often exhibit dynamic patterns. This is particularly true when the product is new and there are not enough field data available. Figure 8.8 is a snapshot of an FM for a length of 13 weeks. It is very common that some dominant FM may shift to the right side of the chart while other trivial FM could pop up and become dominant failures later on. There are several reasons to explain this phenomenon: (1) ongoing CA activities eliminating some major FM but could also induce new FM at the same time; (2) latent failures, which did not show up in early time; and (3) inappropriate usage modes at customer sites, which trigger new FMs.

Figure 8.9 shows the FM Pareto chart for the same PCB product line based on field returns in the second quarter. Now the Op-amp becomes the largest FM with 20 failures in total. The relay issue remains the same as it did in the first quarter. The capacitor issue is shifted to the right and becomes a nondominant FM. The quantity of capacitor failures, however, is five, which is the same as that of the first quarter. Does this mean the CA on the capacitor is not effective or the failure rate remains the same as that in first quarter? Also, the total failure number actually increased from 28 in Q1 to 60 in Q2. Does this imply the product MTBF is becoming worse in the second quarter? Finally, given a changing FM Pareto chart like Figure 8.9, should we continue to implement the CA against capacitor issue? Or do we need to prioritize the CA against Op-amp failures now? What is the MTBF incremental after eliminating critical FM? Obviously these questions cannot be answered by the traditional FM Pareto chart. In the next section, a read-time FM run chart will be proposed to address these issues.

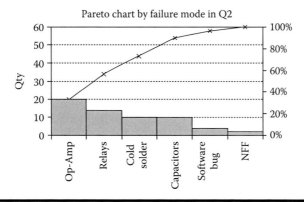

Figure 8.9   Failure mode Pareto chart in the second quarter.

8.4.2 Real-Time FMR Chart

Pareto charts represent the distribution of the product FM for a period of time such as one quarter. In order to evaluate the effectiveness of the CA, one needs to compare several Pareto charts in a consecutive sequence. Moreover, the chart cannot reflect the change of the actual number of failures due to the increase of the field products. When product installations are expanding, more field returns are anticipated during the same period of time even if the product reliability remains the same. Therefore, it is imperative to propose a new metric that is able to identify the dominant FM, but also can forecast the FM trends in the context of the growing field products. To meet both criteria, Jin and Liao (2007) introduced a new FM metric called failure mode rate (FMR), which is defined as

$$FMR = \frac{\text{failures for a type of FM}}{\text{field product population}} \tag{8.14}$$

The FMR effectively integrates the actual FM quantity with field installation of the product. Field installation represents the total shipment of the product or total population of the products. For example, assume 80 PCB modules were installed in the first quarter. Among them five field failures were found due to bad relays, then the FMR for bad relay in the first quarter can be estimated as

$$FMR = \frac{5}{80} = 0.062 \, \text{faults/board} \tag{8.15}$$

In the second quarter, if additional 120 PCB were installed in the field, this would make the field PCB population reach 200. Because of the ongoing CA, the total relay failures in the second quarter remain five despite the increase of the product population. Then the FMR for bad relays in the second quarter can be estimated as

$$FMR = \frac{5}{200} = 0.025 \, \text{faults/board} \tag{8.16}$$

It is noticed that the FMR of relays in the second quarter is reduced by 60% compared to the first quarter although the total failure quantity remains the same. In this example, FMR is computed on a quarterly basis, meaning product failures are averaged in a 13 week window. Since most companies generate quarterly reports using the 13 week window, we used this criterion here to compute the FMR. The definition in Equation 8.14 is very general and FMR can be estimated using any length of time period based on the product failure characteristics and the shipment rate.

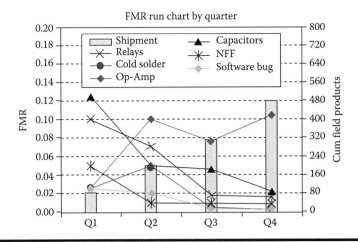

Figure 8.10 Failure mode rate run chart.

Figure 8.10 shows the FMR run chart for four consecutive quarters or 1 year period. The horizontal axis represents time and the vertical axis on the left represents FMR. The cumulative product population is shown on the right vertical axis. This run chart clearly indicates that failures of relays, resistors, and cold solder joints are getting improved through third and fourth quarters. At the end of the fourth quarter, only the Op-amp remains as an outstanding issue on which more efforts need to be focused upon.

In the FMR run chart, the trend of FM can be easily determined, based on which a decision can be made whether CA should be implemented or not. For instance, it is obvious that more attention should be paid to bad Op-amp as it increases from 0.021 in Q1 to 0.1 in Q2. The FMR run chart can also be used to examine the effectiveness of the ongoing CAs. For instance, a CA was implemented for capacitors in the Q1 and the FM rate decreased from 0.12 to 0.02 at the end of Q4. This implies the CA was very effective as the rate was decreased by 600%. Similar conclusions can be made from failures for relays and cold solder joints.

8.4.3 Other Advantages of FMR

The FMR run chart in Figure 8.10 contains important data necessary for decision makers to evaluate product reliability and the effectiveness of ongoing CA. In the industry, when reports are sent to the upper level management, quite often detailed data are often visualized by charts to facilitate the decision-making process. If the management wants to recapture the actual number of failures for a particular FM, the FMR chart is capable of retrieving the actual failures for one specific FM. For example, in Q3, the FMR for capacitor is 0.045 and the cumulative field PCB

Table 8.2 Product 13 Week Rolling MTBF Based on FMR Run Chart

Quarters	Q1	Q2	Q3	Q4
Cumulative product installed	80	200	310	480
Cumulative operation hours	174,720	436,800	677,040	1,048,320
Cumulative FMR	0.350	0.300	0.151	0.154
Cumulative failures	28	60	47	74
MTBF (h)	6,240	7,280	14,424	14,174

population is 310. The number of boards returned due to defective capacitors can be quickly estimated as $0.045 \times 310 = 14$. Similarly, we can estimate the actual failures for other types of FM. Given the cumulative field failures per quarter, the management can allocate adequate resources in the repair center and forecast the spare provisioning for field replacement.

The product MTBF can be estimated implicitly from the FMR chart as well. This is perhaps one of the greatest benefits of using the run chart. The 13-week rolling MTBF can be easily inferred from the FMR chart. For example, in Q1, the cumulative FMR is the summation of all individual FMR as follows:

$$0.125 + 0.1 + 0.05 + 0.025 + 0.025 + 0.025 = 0.35 (\text{faults/board}) \qquad (8.17)$$

Notice that 80 products are installed in Q1, so the total number of failures is $80 \times 0.35 = 28$. In the semiconductor industry, the production is typically running on a 24×7 basis, on all days of the week. The operating hours of each board is 2184 h in 13 weeks. The total cumulative run hours for 80 boards is $2,184 \times 80 = 174,720$ h. Therefore the 13 week rolling MTBF is $174,720/28 = 6,240$ h. Table 8.2 summarizes MTBF values for the PCB board in four different quarters based on the FMR chart in Figure 8.10.

8.5 CA Effectiveness Function

The effectiveness function aims to link the percentage of failure reduction with the amount of the budget used for the CA. A quantitative model will be proposed to evaluate the CA effectiveness in terms of percentage of FM eliminations. The maximum CA effectiveness is 1 if a particular FM is completely eliminated from field systems, and the minimum is 0 if no CA is applied. Typical effectiveness values vary between 0 and 1 depending on the CA methods. More specifically, CA

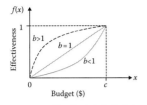

Figure 8.11 CA effectiveness functions.

effectiveness depends on the amount of resources (e.g., budget) allocated for the CA process. Given a specific FM, it is reasonable to assume that the more the budget is allocated for CA, the higher the effectiveness will be. For instance, retrofit can be applied in lieu of ECO if sufficient budget is available to allocate spare parts to replace problem modules. To that end, the following model is proposed:

$$f(x) = \left(\frac{x}{c}\right)^b \qquad (8.18)$$

where x is the amount of the budget for CA. Notice that b and c are parameters and both of them are positive numbers. These parameters can be estimated based on historical CA data or from predecessor systems. The value of c actually is equal to the retrofit cost assuming all field systems receive retrofit service. It can also be easily inferred from the fact that when $x = c$, $f(x) = 1$. Therefore parameter c can be determined based on the amount of money needed, assuming that the retrofit is applied against the FM. b is the shape parameter that controls the curve of the effectiveness function. Depending on the value of b, three types of effectiveness functions are available. Figure 8.11 depicts the effectiveness function for different b.

8.5.1 Linear Effectiveness Model

When $b = 1$, the general model is simplified as a linear function. The linear model indicates that the CA effectiveness is proportional to the amount of budget spent on the FM elimination:

$$f(x) = \frac{x}{c} \qquad (8.19)$$

This model is relatively simple, yet it has wide applications due to its mathematical convenience and simplicity. For many practical problems, the actual CA effectiveness can be approximated by the linear model.

8.5.2 Power Effectiveness Model

If $b > 1$, $f(t)$ becomes a power function. For example, if $b = 2$, the following quadratic effectiveness function is obtained:

$$f(x) = \left(\frac{x}{c}\right)^2 \qquad (8.20)$$

The quadratic model represents the situation where the CA effectiveness is small if a small amount of budget is allocated. However, the effectiveness increases quickly if the CA budget exceeds a certain quantity (see Figure 8.11). For instance, a device lifetime can be extended by installing a heat sink as a containment approach (which is cost effective). Yet the device lifetime will not be significantly improved unless the heat is significantly reduced through redesign. The latter requires much more engineering resource.

8.5.3 Rational Effectiveness Model

The rational model represents the effectiveness when the parameter $b<1$ (e.g., $b=1/2$).

$$f(x) = \left(\frac{x}{c}\right)^{\frac{1}{2}}$$
(8.21)

The rational model represents the situation where the CA is very effective just if a small amount of money is allocated. However, the effectiveness decreases once the CA budget reaches certain amount of money. Typical examples include software upgrade to remove known high-frequency bugs from existing software applications.

8.6 Reliability Growth Monitoring

Ideally, the product is not recommended for volume manufacturing before component and noncomponent failure rates reach the target value. Driven by the fast technology cycle in the semiconductor industry, new test equipment is often pushed to the volume production even if component and noncomponent failure rates are still at a high level. In other words, product reliability often yields to the shipment deadline in order to gain the competitive market share. Therefore RGP and CA are extended from the product development, pilot line production, to the early volume shipment until the target MTBF is obtained. Hence, it is necessary to develop an effective monitoring system to track the reliability growth of the product during such an extended time period. The system is also able to quantitatively pinpoint who is the key stakeholder for component and noncomponent issues.

8.6.1 FIT-Based Monitoring System

In the product design phase, the target MTBF is often determined based on BOM, customer requirements, and competitor's performance. As shown in Table 8.1, noncomponent failures could count up to 50% of all failures in the product's early volume shipment. In Section 8.3, we have presented a stochastic-based reliability prediction model that explicitly incorporates noncomponent failures. Given the system failure rate, the system MTBF can be estimated by

$$MTBF = \frac{1}{\lambda},$$ (8.22)

where λ is the product or system failure rate. However, the system failure rate can also be expressed as failures in time or FIT, that is, the number of failures in 10^9 h. The relationship between the failure rate and the FIT quantity (failures in 10^9 h) is given by

$$FIT = \lambda \times 10^9 = \frac{10^9}{MTBF}$$ (8.23)

For example, if a current system's MTBF is 1000 h (or $\lambda = 0.001$ failures/h), then the system's FIT is 10^6, meaning the system will fail one million times if it operates for one billion hours. It is worth of mentioning that λ could be used to describe the failure rate of system, or the failure rate of components and noncomponents. For the latter, values estimated from Equation 8.23 stands for component or noncomponent FIT. The concept of FIT could be further extended to a specific noncomponent such as manufacturing, design, software, process, and others.

8.6.2 Illustrative Example

We used the example from Reference [19] to demonstrate the application of the FIT-based monitoring system. A PCB board with the target MTBF of 50,000 h is required after 1 year of field shipment. During the concept and design phases, engineering team came up with a MTBF budget table that breaks down the product MTBF into six categories as shown in Table 8.3.

The target MTBF for each category is estimated based on existing product reliability and the complexity of the new product. Given the target MTBF of each category, the corresponding FIT rate is also computed using Equation 8.23 and listed at the right column. The cumulative FIT is 20,000, which is equivalent to 50,000 h target MTBF. The advantage of FIT over MTBF is the capability of the summation of individual FM category in estimating the product MTBF.

Table 8.3 further breaks down the FIT value based on the individual FM with the target value. Each FM is assigned to a person who is responsible for the reduction of the FIT rate. If the actual FIT rate is bellow the target, no CA is needed. Those FM creating the largest MTBF gap usually is prioritized for CA, and decision can be made together with the stakeholder listed in the last column of the table.

Table 8.3   Convert Target MTBF into FIT Values

FM Category	Target MTBF	Target FIT	Stakeholder
Components (hardware)	117,647	8,500	John
Others (NFF)	250,000	4,000	Mike
Design	333,333	3,000	Jason
Manufacturing	500,000	2,000	Josh
Process	666,667	1,500	Dave
Software	1,000,000	1,000	Jose
Board (total)	50,000	20,000	William

8.7 Conclusion

This chapter proposes several new techniques in managing product reliability in the distributed manufacturing environment. These techniques include: (1) a stochastic-based reliability prediction model; (2) real-time FMR run charts; (3) CA effectiveness functions; and (4) a FIT-based reliability monitoring system. These new models aim to provide a realistic reliability prediction by accommodating both component and noncomponent failures during the product design phase. The FMR chart was developed to dynamically track the FM trend, CA activities, and the incremental of the field products. The CA effective function is able to guide the decision makers to prioritize CA resources in order to achieve better reliability growth. The FIT-based reliability monitoring system provides the engineering team the most efficient information in identifying the gap between the target reliability and the current reliability in terms of the component, design, software, and manufacturing etc. As such, each member of the engineering team exactly knows her/his responsibilities, which motivates their involvement in RGP.

As the electronic industry continues to shrink the product development cycle, new products are often pushed to volume manufacturing before all significant FM are removed. Coupled with increased shipment quantity, CA activities are extended from in-house design to volume shipment. Reliability must be embedded into the early design phase if the manufacturer wants to maintain competitiveness in the global market. Both component and noncomponent risks need to be mitigated by identifying and eliminating high potential risks. As the product is ramped up to volume shipment, field failure data should be systematically tracked and recorded, which will serve as the basis for CA decisions and implementations. These measures lead to enhanced customer satisfaction and more products will be purchased by both existing and new customers.

Authors

Tongdan Jin is an assistant professor in the Ingram School of Engineering at Texas State University. He was an assistant professor of systems engineering at Texas A&M International University (TAMIU) between 2006 and spring 2009. Before A&M, he held a reliability engineer position at Teradyne Inc., Boston, for 5 years. He received his PhD in industrial and systems engineering and his MS in electrical and computer engineering from Rutgers University. He has published more than 40 articles in various journals and conference proceedings. His research interests include system reliability modeling and optimizations, process modeling for micro/nanomanufacturing, and virtual energy provisioning for smart grids.

Madhu Kilari is a graduate student in the Department of Business Administration at TAMIU. Between 2008 and 2009, he worked as a research assistant in the Department of Engineering, Mathematics and Physics under the supervision of Dr. Tongdan Jin. He received BS in information and communication technology from Dhirubhai Ambani Institute of Information and Communication Technology, India. He worked as a software engineer for 2 years in Infosys before joining TAMIU in 2008. He has experience in using C/C++, MATLAB, and Cplex programs.

References

Chen, Y.K. and Sackett, P.J. 2007. Return merchandize authorization stakeholders and customer requirements management of high-technology products. *International Journal of Production Research* 45(7): 1595–1608.

Clark, J.A. 1999. Modeling reliability growth late in development. In *Annual Reliability and Maintainability Symposium*, Washington, DC, pp. 201–207.

Crow, L.H. 1974. Reliability analysis for complex, repairable systems. In *Reliability and Biometry*, Philadelphia, PA, SIAM, 379–10.

Crow, L.H. 1984. Methods for assessing reliability growth potential. In *Proceedings of Reliability and Maintainability Symposium*, Piscataway, NJ, pp. 484–489.

Crow, L.H. 2004. An extended reliability growth model for managing and assessing corrective actions. In *Annual Reliability and Maintainability Symposium*, Los Angeles, CA. 73–80.

Duane, J.T. 1964. Learning curve approach to reliability monitoring. *IEEE Transactions on Aerospace* 2: 563–566.

Ellner, P.M. and Hall, J.B. 2006. An approach to reliability growth planning based on failure mode discovery and correction using AMSAA projection methodology. In *Proceedings of Reliability and Maintainability Symposium*, Newport Beach, CA, pp. 266–272.

Gullo, L. 1999. In-service reliability assessment and top-down approach provides alternative reliability prediction method. In *Annual Reliability and Maintainability Symposium*, Washington, DC, pp. 365–377.

Jin, T. and Liao, H. 2007. Failure time based reliability growth in product development and manufacturing. In *Annual Reliability and Maintainability Symposium*, Orlando, FL, pp. 489–493.

Jin, T., Liao, H., and Kilari, M. 2010. Reliability growth modeling for in-service systems considering latent failure modes. *Microelectronics Reliability* 50(3): 324–331.

Jin, T., Wang, P., and Huang, Q. 2006. A practical MTBF estimate for PCB design considering component and non-component failures. In: *Annual Reliability and Maintainability Symposium*, Newport Beach, CA, pp. 604–610.

Johnson, B. and Gullo, L. 2000. Improvements in reliability assessment and prediction methods. In *Annual Reliability and Maintainability Symposium*, Los Angeles, CA, pp. 181–187.

Krasich, M. 2006. Accelerated reliability growth testing and data analysis method. In *Proceedings of Reliability and Maintainability Symposium*, Newport Beach, CA, pp. 385–391.

Krasich, M., Quigley, J., and Walls, L. 2004. Modeling reliability growth in the product design process. In *Annual Reliability and Maintainability Symposium*, Los Angeles, CA, pp. 430–442.

Smith, T.C. 2004. Reliability growth planning under performance based logistics. In *Proceedings of Reliability and Maintainability Symposium*, Los Angeles, CA, pp. 418–423.

Telcordia Technologies. 2001. Reliability prediction procedure for electronic equipment (issue 1) special report TR-332. Telcordia Customer Service, Piscataway, NJ.

US MIL-HDBK-217. 1965. Reliability prediction of electronic equipment (version A).

Wang, P. and Coit, D.W. 2005. Repairable systems reliability trend tests and evaluation. In *Proceedings of Annual Reliability and Maintainability Symposium*, Alexandria, VA, pp. 416–421.

http://www.brightonwebs.co.uk/distributions/triangular.asp

Chapter 9

Taguchi Integrated Real-Time Optimization for Product Platform Planning: A Case of Mountain Bike Design

Mukul Tripathi and Hung-da Wan

Contents

In today's competitive market, the explosion of product variety is driving the focus of firms from mass production to mass customization. Mass customization aims to satisfy individual customer needs at a price comparable to those of mass produced goods. The fulfillment of customization needs through individualized products has commonly been approached through product family planning. The concept of product family has been recognized as an effective means for supporting variety by sharing common features, components, and subsystems yet satisfying diverse market niches. This chapter proposes a new integrative approach to the allocation of adjustability and sizing of mountain bike frame. The case study undertaken constitutes a problem of industrial strength with numerous design variables. The underlying platform planning problem is solved utilizing a novel optimization methodology entitled as Taguchi integrated real-time optimization (TIRO). In order to show the robustness of the model to handle various kinds of market fluctuations, various demand modeling techniques have been detailed and included into the mathematical model. Moreover, the efficacy and the supremacy of the proposed TIRO solution procedure are benchmarked against two sets of test beds that were meticulously generated by utilizing techniques derived from design of experiments (DOE) and various demand modeling functions. Analysis of variance (ANOVA) is performed to verify the robustness of the proposed solution methodology. The algorithm performance has also been compared with three other pure algorithms (genetic algorithm [GA], age-GA, and sexual-GA) where TIRO was seen to outperform significantly in all the problem instances within the test bed.

9.1 Introduction

In today's competitive and highly volatile market, it becomes a key issue for companies to best meet diverse demands of customers by providing variety of products in a timely and cost-effective manner. As a response to the explosion in product variety, the companies are redefining their manufacturing activities by adopting slogans such as "better," "faster," and "cheaper" (Simpson, 1998). "Customers can no longer be lumped together in a huge homogeneous market, but are individuals whose individual wants and needs can be ascertained and fulfilled" (Pine, 1993). Therefore, the new customer-driven market is shifting the focus of firms from mass production to mass customization by adopting a redesign of their existing products. Mass customization aims to satisfy the individual customer needs at a price comparable to those of the mass-produced goods. The fulfillment of customization needs through individualized products has commonly been approached through product family planning. The concept of product family has been recognized as an effective means for supporting variety by sharing common features, components, and subsystems yet satisfying diverse market niches. With growing concern for satisfying individual costumers, companies now face the challenge of providing as much variety as possible for the market while having maximum commonality between the products for the producer.

In order to consider the market benefits of customization and cost of providing variety, generally product platform sharing common components and subsystems are built up. The product platform acts as an interface for augmentation, modification, and renewal to form a product family, thereby resulting in the generation of derivative products. A product family is defined as the set of individual products that share common technology and address a related set of market applications (Mayer and Lehnerd, 1997). Platforming has recently received increased attention with regard to its wide utility in product development. Thus, one of the major successes for burgeoning of mass customization is determining commonality in platform design variables.

Against this background, the approach advocated in this research is to design and develop a family of products in a view to foster a better understanding of product family concepts in real-world situations of nonuniform demands.

This research investigates the problem of offering a range of customized product variants for markets characterized by nonuniform market demand. This chapter first presents a theoretical foundation for designing product platform as a problem of access in geometric space where all available information is complete and certain, and then utilizes a new encoding scheme to help identify which platform design variables should be considered common. This chapter introduces a mountain bike (MTB) frame fabrication problem as an example for solving product platform design. The system framework is extracted from an MTB production scenario where customer satisfaction is mapped with the *stand-over height* and *reach* of the nearest product variants available to the customer. The proposed system framework is governed by the following three factors: (1) selection of the frame material, (2) the market space and demand scenarios, and (3) weight given to the customer unsatisfaction level measured in terms of the fuzzy linguistic variables. These factors were first varied at different levels in order to carry out an exhaustive experimentation by generating two sets of test beds, and then a novel approach is proposed for determining the optimal platforms in a fixed market space as explained later.

Due to high computational complexity, the platform planning problems fall into the category of NP-hard Problem (i.e., non-deterministic polynomial-time hard problem). Larger instances of such problems cannot be solved by utilizing the traditional deterministic approaches (Tripathi et al., 2008, 2009b). In this research, the authors propose a Taguchi integrated real-time optimization (TIRO) strategy for attaining optimal/near-optimal results in a required timeframe. The TIRO strategy presents an adaptive method for allocation of computational resources among a set of algorithms to achieve a superior performance than pure algorithms (Tripathi et al., 2009a). The approach is "real-time" in the sense that the decision of algorithm selection is made according to the performance of the algorithms during the runtime. The methodology does not rely upon any complex prediction model (either on problem domain or on algorithm behavior) and is capable of achieving a superior performance than pure algorithms. In addition, TIRO adopts two fundamental concepts from design of experiments (DOE), namely, orthogonal arrays (OAs) and signal-to-noise ratio (SNRs) (Phadke, 1989). The principal idea is to utilize the robust traits of statistical experimental design to enhance the solution quality by utilizing only a partial set of experiments without neglecting the cause of variation of the factors. Motivated by the encouraging results reported in the literature (Tsai et al., 2004, 2006), the authors implemented the robust DOE principles in the optimization algorithm. The underlying principle involved in the process is based on maximizing the performance measures (i.e., SNRs) by conducting only a reduced set of experiments using OAs. For validating the efficacy and efficiency of the proposed TIRO technique, a test bed is generated by selecting seven factors and varying them at two levels over which analysis of variance (ANOVA) is performed. The robustness of the proposed algorithm is authenticated against the individual pure algorithms undertaken in the TIRO approach.

The rest of the chapter is organized as follows: Section 9.2 contains the literature review of problem domain and the solution methodology, while Sections 9.3 and 9.4 discuss the product platform planning problem and then mathematically formulate it, taking the case study of an MTB frame fabrication. In the next section, the detailed functioning of the TIRO strategy is delineated. Computational results and insights are developed in Section 9.6, and finally Section 9.7 concludes the chapter with suggested directions for study and research.

9.2 Literature Review

A product family is defined as a set of individual products that share common technology and address a related set of market applications (Mayer and Lehnerd,

1997). The key to design an effective product family is platform-based product development where platform variants are derived by augmenting the common core of the product in one or more dimensions to target specific market niches (Simpson, 2004). In general, platform-based mass customization practices have been grouped into two broad classifications, namely, module-based product families and scale-based serialization of product families (Simpson, 2005). In module-based product families, the common core or the platform is augmented by adding, substituting, removing, or sharing one or more functional elements or modules. Nanda et al. (2005) use the formal concept analysis (FCA) to represent and analyze the product family consisting of four cameras with specific functionalities. Thevenot et al. (2006) undertake a case study based on real data and discussions from a company producing a line of innovative staplers with a limited set of attributes. In contrast to modularity, the scale-based product family derives its constituting elements by scaling one or more design variables. Several works on scale-based serialization existing in literature have been classified into various benchmark problems by Allada et al. (2006). The product platform design problem with pressure vessel as a case study was studied by Hernandez (2001). They utilized product platform constructal theory method (PPCTM) to solve the problem via deterministic methods. Williams (2003) and Williams et al. (2004) undertake the same problem and modeled demand scenarios as being nonuniform in nature and attacked the problem using an augmented PPCTM approach.

In recent times, there has been extensive use of artificial intelligence techniques, such as simulated annealing, genetic algorithm (GA), Tabu search, etc., for the optimization of computationally complex problems. These algorithms are marked by their short response times and high-quality solutions over the problems of real dimensions. The algorithm selection problem consists of choosing the best algorithm from a predefined set to run on the given problem instance (Rice, 1976). The problem has been utilized by many researchers using both high knowledge classification approach and low knowledge control of optimization algorithms. Lagoudakis and Littman (2000) used Markov decision process with reinforcement learning applied to algorithm control. Boyan and Moore (2000) attempted to correlate problem features with search performances in an effort to improve the searching procedure. Carchare and Beck (2005) applied machine learning approach and introduced term "low-knowledge control" for optimizing scheduling problems. In this research, the authors propose relative improvement factor (RIF) for each pure algorithm during runtime and thereby propose the TIRO search strategy for determining the optimal/near-optimal solution of the product platform planning problem.

9.3 Mountain Bike Frame Design

In general, bicycle manufacturing includes three main activities: frame design and parts selection, frame fabrication, and final assembly. In this chapter, the authors

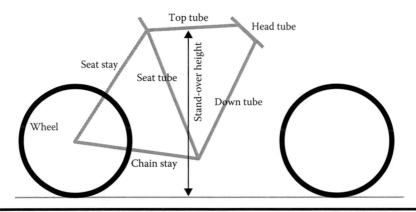

Figure 9.1   The technical details of a bicycle frame.

focus on the first part, i.e., the frame design, and present a case of platform planning for a bicycle manufacturing firm. For the design and manufacturing of a bicycle frame, the most important part is the diamond-shaped frame, which is responsible for strength and rigidity of the bicycle, keeping the components together in proper geometric configuration. In general, the frame consists of a front and a rear triangle as shown in Figure 9.1. The front triangle is, in turn, composed of four tubes, namely, (1) top tube, (2) down tube, (3) seat tube, and (4) head tube. Similarly, the rear triangle consists of following two parts: (5) chain stay and (6) seat stay. Attached to the head tube at the front is the steering assembly, including the fork, handlebar, etc. Other components of the bicycles, such as wheels, derailleur, brakes, and chains, are usually purchased by the bicycle assembler. Figure 9.1 illustrates the technical details of a bicycle frame concerned in this research. In this case study, the bicycle frames are made of metal tubes with elliptical cross section defined by a horizontal and a vertical diameter and by the thickness of the tube. Other parameters utilized in the MTB frame fabrication are detailed in the Appendix 9.A.

9.4 Modeling the Design Problem

In this research, the problem of offering a range of customized product variants with nonuniform market demand is investigated. A case study of MTB frame fabrication with nonuniform demands throughout a static market domain is presented, with an attempt to maximize the average profit of all the product variants in the entire market space. This market space or the product offering domain (POD) is formed by setting bounds over two parameters used commonly by the manufacturers to roughly convey the bicycle frame geometry information to their customers. By providing the customers the opportunity to select bicycle frames designed for their specific needs, it is assumed that the resulting customized product family will be of high demand and therefore profitable.

9.4.1 Assumptions

There are three key assumptions in formulation of customizable MTB frames:

1. It has been assumed that the market space and the demand scenarios do not expand or contract over the considered timeframe. Following this assumption, the product platform problem has been formulated for a fixed market space.
2. The product variants are assumed to possess fixed number of functionalities and the manufacturer can develop the product variant only by scaling the product components.
3. The model formulation does not account for uncertainties and risk inherent either in design variables or in the demand model.

9.4.2 Mathematical Model

The manufacturer seeks a competitive advantage over other leading manufacturing enterprises by offering customized product variants to the potential customers. The target customers for MTB frame manufacturing firm are the professional riders who are so much attuned to their riding needs that they prefer bikes, which are customized to fit them best. In general, two body dimensions are selected by them as performance parameters to determine the acceptable bicycle size: stand-over height (h_{SO}) and *reach* (I_R). The manufacturer tends to minimize the customer unsatisfaction level by matching the exact performance requirements. These two performance measures are mapped with the design variables for the formulation of mathematical model. The stand-over height is directly related to the geometry of bike via its three design parameters, i.e., seat tube angle (θ_s), seat tube length ($L_{ST_{cc}}$), and height of the crank set (h_{cs}) and generally is a measure adopted by the manufacturer to name its variants.

$$h_{SO} = h_{cs} + L_{ST_{cc}} \times \sin(\theta_s) \tag{9.1}$$

The second parameter being mapped with the design variable is the reach of the rider associated with the effective top tube length ($L_{TT_{eff}}$) of the bike.

$$l_R = L_{TT_{eff}} + 6 \tag{9.2}$$

For each of the demand scenarios, the values of the following design variables are to be calculated by utilizing Equations 9.1 and 9.2 and considering the related technical constraints:

1. Head angle (θ_h)
2. Seat angle (θ_s)

3. Radius of the wheels (r_w)
4. Lengths of the six tubes involved in defining the frame geometry

We state the objective function as the maximization of the average profit of all the product variants. Mathematically, we have

$$\text{Avg Profit} = \left(\sum_{i=I_1}^{I_2}\sum_{j=J_1}^{J_2} D_{ij}\right)^{-1} \times \left(\sum_{i=I_1}^{I_2}\sum_{j=J_1}^{J_2}(\text{Net Profit})_{ij}\right) \tag{9.3}$$

where the subscript $i \in (I_1, I_2)$ refers to the stand-over height and $j \in (J_1, J_2)$ defines the reach of MTB frame. The potential profit from each product variant (i, j) is obtained by differencing the selling price and the cost of manufacture of that product variant as shown in Equation 9.4.

$$(\text{Net Profit})_{ij} = SP_{ij} - CP_{ij} - CU \tag{9.4}$$

where total selling price SP_{ij} is set by the manufacturer by defining two cost multipliers CC_{I_i} and CC_{J_j} with negative gradients. Figure 9.2 represents the variation in CC_{I_i} and CC_{J_j}. The manufacturer rates the bicycle more by the stand-over height criteria, which is clearly visible by the dominance of CC_{I_i} over CC_{J_j} in the figure.

$$SP_{\text{tot}} = \sum_{i=I_1}^{I_2}\sum_{j=J_1}^{J_2} D_{ij} \times \left\{\left(CC_{I_i} \times I_i + CC_{J_j} \times J_j\right)\right\} \tag{9.5}$$

$$CP_{\text{tot}} = \sum_{i=I_1}^{I_2}\sum_{j=J_1}^{J_2} D_{ij} \times \left(C_{\text{metal}} \times W_{\text{metal}} + C_{\text{weld}} \times N_{\text{weld}}\right) \tag{9.6}$$

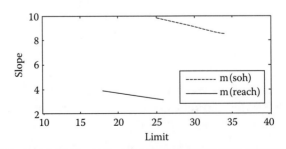

Figure 9.2   Represents the variation in CC_{I_i} and CC_{J_j}.

$$\sum_{i=I_1}^{I_2}\sum_{j=J_1}^{J_2}\left(\text{Net Profit}\right)_{ij} = \sum_{i=I_1}^{I_2}\sum_{j=J_1}^{J_2}D_{ij}\times\left\{\begin{array}{l}\left(CC_{I_i}\times I_i + CC_{J_j}\times J_j\right)\\-\left(C_{\text{metal}}\times W_{\text{metal}} + C_{\text{weld}}\times N_{\text{weld}}\right)\\-\left(C_{\text{cul}}\times cul\right)\end{array}\right\} \quad (9.7)$$

$$-\,C_{\text{order}} - C_{\text{equip}}$$

A product variant is represented as a pair (i, j), which varies from (I_1, J_1) to (I_2, J_2) to cover the entire market space. D_{ij} is the demand of a specific product variant (i, j), SP_{tot} is the total selling price of all the product variants and CP_{tot} is its corresponding cost of manufacture. C_{order} and C_{equip} are the total ordering cost and the setup costs for the equipments, respectively, and depend upon the total number of product variants generated. Average profit and net profit of all the product variants is calculated as per the demand of that variant as shown in Equations 9.5 and 9.7, respectively. Cost price contains a term W_{metal}, which represents the weight of the metal and is calculated by defining the geometry of the bicycle and is calculated in Appendix 9.B.

The customer unsatisfaction or CU for that product variant in the entire market space is calculated based on its proximity with the actual product variants offered by the manufacturer. Keeping in mind the vague nature of unsatisfaction expressed by the customers, the fuzzy set theory is employed to approximate the customer unsatisfaction in terms of fuzzy linguistic variables. Consider a customer who demands product variant (i, j) but contends with a variant (i_x, j_x) produced by the manufacturer. We calculate two intermediate variables:

$$\Delta i_x = |i - i_x| \quad (9.8)$$

$$\Delta j_x = |j - j_x| \quad (9.9)$$

where $x \in \{1, 2, \ldots, N\}$ and N is the number of product variants offered by the manufacturer. These Δi_x and Δj_x are passed through a fuzzy logic controller (FLC) with multiple input single output (MISO) structure with two inputs and one output. The authors conducted a small survey to find out customer reaction and sensitivity toward stand-over height and reach of the bicycle. Based on this, we frame the rule base for our FLC as shown in Table 9.1.

We adopted the triangular-shaped input and output membership functions with 25% overlapping of three linguistic variables (low, medium, and high) in all the cases. Further defuzzification was carried out by utilizing centroid method (Ross, 1997). These Δi_x and Δj_x serve as inputs for the FLC and returns the value of customer unsatisfaction level (cul_x). The minimum cul_x corresponds to the best

Table 9.1 Rule Base Providing Fuzzy Expert Rules

	Reach		
Stand-over Height	Low	Medium	High
Low	1	1	2
Medium	2	2	3
High	2	3	3

Note: 1, = less; 2, medium; 3, more.

product variant available to the customer and is thereby selected as cul_x for the current product under consideration.

$$cul = \text{Min}\{cul_1, cul_2, ..., cul_N\} \qquad (9.10)$$

This *cul* is multiplied with its associated cost constant for (C_{cul}) to obtain the corresponding deficit in profit of the manufacturer.

$$CU = C_{cul} \times cul \qquad (9.11)$$

The effect of C_{cul} on the objective function value is studied in the ANOVA analysis in the experimental design and results section.

9.4.3 Market Description

The basic requirement for the manufacturer is to produce a range of customized product variants where the limits of stand-over height and reach of the bicycle have been explicitly defined, thereby making the market space fixed. This POD or the market space for bivariate normal distribution is shown in Figure 9.3.

In order for the products to perform favorably in the market, it is imperative that the manufacturer is familiar with the market realities and trends. Hence, as a prerequisite to effective platform planning, the overall view of the market demand scenarios should be available to the base decision makers. While the size of the market space is uniform over time, the demand for individual product variants within the market is not. In general, the demand is found to be nonuniform across the entire market space as the uniform demand approximation does not adequately capture the system complexity associated with the traditional market for a customized production.

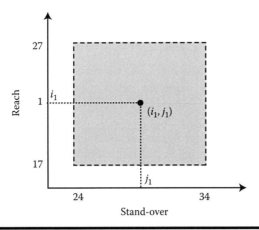

Figure 9.3 **The POD or the market space for bivariate normal distribution.**

The technical constraints are as follows:

1. Limits on the range of the wheel radius constrained by the source of the supply to the manufacturer, i.e., $285 \leq r_w \leq 330$.
2. Constraints applicable on the design variables governing the geometry of the frame as provided in Appendix 9.A.
3. Constraints on seat tube and top tube in the current POD defined by Equations 9.1 and 9.2.
4. Limitation on the range of head angle ($65 \leq \theta_h \leq 76$) and seat angle ($68 \leq \theta_s \leq 73$) for feasible frame structure.

9.5 Solution Methodology

The product platform planning problem is computationally complex in nature and belongs to the class of NP-Hard problems. Traditional methods of attaining global maximum in a multimodal function with unknown number of maxima rely generally on the stochastic search techniques (Mandal et al., 2007). However, significant variation among the final outputs produced by these techniques is evident by the greater standard deviation in the results generated by the same algorithm with different random seeds (Tripathi et al., 2009a). This may sometimes lead to inefficacy of the random search technique by producing results entrapped in the local maxima (Shukla et al., 2009). Hence, certain measures must be incorporated into the solution methodology, which makes it capable of generating results with acceptable standard deviation (Nagalakshmi et al., 2009). In this chapter, we propose an optimization framework comprising of a set of stochastic search algorithms integrated with the concepts of DOE. We observed following two characteristics of the proposed methodology:

1. Producing results with greater proximity toward global optimum than the pure algorithms under consideration.
2. Producing results with smaller standard deviation.

The computational complexity of the aforementioned problem paves the way for development of a search technique that efficiently predicts and selects a better algorithm from a given set and adequately explores the entire search space. A schematic representation of the algorithm flow is given in Figure 9.4. From the same figure, two phases of the algorithm are clearly visible: an algorithm selection and control phase and a Taguchi selection phase.

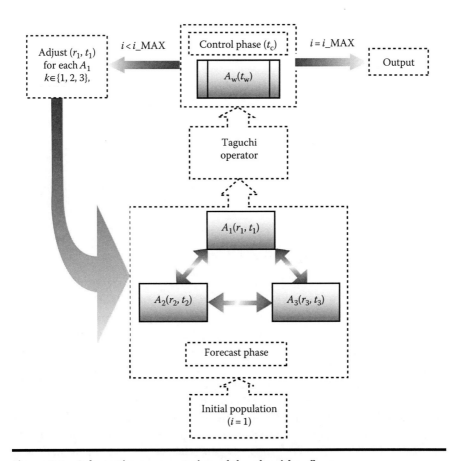

Figure 9.4 Schematic representation of the algorithm flow.

9.5.1 Algorithm Selection and Control Phase

In general, superiority of a search strategy is judged by its relative performance over other metaheuristics by the principle of winner takes all (Rice, 1976). However, an algorithm producing better results on one problem instance may not guarantee to produce similar results in all other cases (Kumar et al., 2009a). Therefore, any single optimization strategy may not prove itself to be versatile enough for having universal applicability in terms of generating better averaged results. Hence, the algorithm selection and control poses an issue of significant relevance in current optimization technology. In view of above considerations, the authors propose a TIRO search strategy as an adaptive method for allocation of computational resources among a set of algorithms to achieve a superior performance on the product platform problem with variable customer demand. The approach followed in this chapter does not rely upon any complex prediction model (either on problem domain or on algorithm behavior) and performs iteration wise selection of the algorithms (Kumar et al., 2009b). In the proposed technique, the metaheuristics compete among themselves for both their selection and control.

Given a time limit T, TIRO ranks each algorithm during runtime in the order in which they have to operate on the problem instance. The time T is further divided into a short forecast phase (t_f) and control phase (t_c) such that

$$T = t_f + t_c \qquad (9.12)$$

The forecast phase (t_f) predicts which algorithm should be utilized for remaining of the control time $T - t_f = t_c$. We represent the information flow within the real-time optimization-based strategy as shown in Figure 9.4. $A_l(r_l, t_l)$, $l \in \{1,2,3\}$, represents an algorithm from the predefines set, r_l and t_l being its corresponding rank and time for which the algorithm runs during the forecast phase. A_w is the winner algorithm that is predicted to perform better in the control phase and t_w is the time for which it is run. It is clear from Figure 9.4 that $t_w = t_c$. The winner algorithm amongst the three is decided by the RIF_l calculated during the forecast phase via a real-time algorithm selection procedure.

The random search techniques respond sensitively to the initial feasible solution presented to them. This is why averaged result (for numerous random seeds) is provided when these stochastic search metaheuristics are applied to any problem instance. The bottom line behind employing the real-time algorithm selection strategy is to forecast which metaheuristic (from a given set) will perform better during the control phase where most of the computational time is allotted to the processor. In order to circumvent the loss of computations carried out during the forecasting phase, the results produced by one algorithm is passed on directly to the other algorithm sequentially. Instead of computing absolute improvement in results for each algorithm over a common static population, a RIF_l is utilized over a dynamic population received by previous algorithm. This RIF_l is defined mathematically according to

$$RIF_l = \begin{cases} 0 & \text{if } \sum_{m=1}^{n} \left(\frac{(ibs_m - bs_m)}{bs_m} \right) = 0 \\[2em] \dfrac{(ibs_l - bs_l)/bs_l}{\sum_{m=1}^{n} \left((ibs_m - bs_m)/bs_m\right)} & \text{otherwise} \end{cases} \tag{9.13}$$

Here, ibs_l is the best solution produced in the current iteration and bs_l is the best solution over which the algorithm operated upon. The runtime for each algorithm is equally distributed (t_f/n) among them during the forecast period. During initialization, the ranks have been randomly assigned to each algorithm which are later updated according to the rank factor rf_k (as shown in Equation 9.14) such that greater the value of rf_k, better the corresponding rank:

$$rf_k = \{n - r_l (i-1)\} \times RIF_l \tag{9.14}$$

where $r_l(i-1)$ represents the rank of the algorithm A_l in the previous iteration. Any conflict in rf_k (arising due to equality of rank factors of two or more algorithms) is broken by randomization of ranks.

As a part of this research, GA and two of its variants, namely, age-GA (AGA) (Ghosh et al., 1996) and sexual-GA (SGA) (Wagner and Affenzeller, 2005), were selected to operate sequentially in each iteration according to their corresponding ranks. Pseudocodes for these three algorithms are presented in Appendix 9.C.

9.5.2 Taguchi Selection Phase

As previously discussed, Taguchi is a robust tool of DOE utilized for optimizing the product and process conditions. In this research, the Taguchi selection is inserted in between the prediction and the control phase. After the prediction phase forecasts the winner algorithm, the population being passed to the control phase undergoes a Taguchi selection procedure. The basic idea of employing the Taguchi technique in our optimization strategy is to study the combination effect of large number of decision variables. However, the beauty of the method lies in the fact that only a partial set of experiments are conducted during this study, thereby saving the expensive computational time involved in the process without neglecting the cause of the variation.

In this research, two fundamental concepts from the Taguchi method are adopted, namely, OAs and SNRs. An OA is denoted as $L_n(2^{n-1})$ and contains "n" experiments as its rows and $(n-1)$ variables (whose effects are to be studied) are taken as its columns. Specifically, we utilize $L_{128}(2^{127})$ for the underlying problem such that at a time, two strings from a population are executed at two different levels for 128 experiments. These strings are selected in pair (each element of the pair is treated as a level of experimentation) from the population based on the best fitness value and the string length to perform the matrix operations in an OA. The length of the string varies dynamically based on the number of product variants. If the number of elements in the string is x at any time, only the first x columns from the OA are selected and the remaining $(128-x)$ columns are ignored in the experimentation. We calculate the objective function value of each factor for each experimental run followed by calculation of SNR or the mean square deviation in the objective function value. Our problem statement considers the maximization of profits and thus SNR for the objective function is calculated as the sum of square of the corresponding individual objective function values $\theta_i = \sum (Obj_i)^2$. Finally, the effect of various factors are denoted by the notation $\alpha_{f,l} = \text{sum of } \theta_i$ for factor f at level l. In our case, $\alpha_{f,1}$ and $\alpha_{f,2}$ are calculated for each factor and compared with each other. If $\alpha_{f,1} > \alpha_{f,2}$, the optimum level is level 1 and vice versa. The iteration is repeated till we cover the complete string length and hence determine the optimum level for each factor in the in the corresponding chromosome string. In this way, a whole set of population is generated for the control phase.

9.6 Experimental Design and Results

This section contains the experimental design utilized and results obtained by implementation of TIRO strategy on the aforementioned problem. The proposed algorithm is coded in C++ and compiled on GNU C++ compiler on a Linux platform. Further, the compiled program is run on a system specification of 2.13 GHz Intel Core2Duo processor and 4 GB of RAM. The algorithm performance is studied on a set of carefully generated test beds. Further, comparisons of performances of proposed approach have been done against the individual pure algorithms involved in the TIRO suite to validate its efficacy over them. We generate two test beds for the performance comparison of TIRO with its individual pure algorithm and for establishing the robustness of the model in demand variations.

9.6.1 Encoding Schema and Parameter Settings

The first step in implementation of a search technique to any problem is the representation of the search space in terms of algorithmic parameters. For current POD, we represent the knowledge-based string as the binary encoding to make the search space continuous. Each string representation is a potential solution to the problem.

The scheme of mapping the search space to the continuous design variables increases the flexibility. However, size of the string is governed by the total number of product variants generated, thus rendering it a data structure with dynamic size.

9.6.2 Test Bed 1 Formulation

The bicycle frame assembly involves numerous design variables, which are structured with complexity. This complexity is resorted by fabricating the frames using tubular components, which not only provide ease of manufacture but also analysis of the same. The problem discussed in this chapter involves numerous factors, out of which, seven factors were identified to have major impact on the results. In an attempt to study their effect on objective function value, these factors were varied at two different levels.

9.6.2.1 Factors

The factors considered for the study are as follows:

1. *Demand range* (D_{ij}): For the test bed formation, the variation in demand is considered to be normally distributed. The demand range was varied at two levels to depict the market with high and low demand scenario. The maximum demand at level 1 (low) demand scenario is kept at 450 and level 2 (high) is kept at 1250.
2. *Frame material density* (ρ_m): Variation in this factor is brought about by changing the frame material itself. The choice of frame material is based on the design criteria, the customer preference, and advancement in materials technology. In selection of the frame material, the customer makes a trade-off between the strength, rigidity, durability, and lightness. We experimented with aluminum at level 1 and steel alloy at level 2.
3. *CUL weightage* (C_{cul}): The effect of the customer unsatisfaction level set by the manufacturer is kept at 150 for low and 200 for high level.
4. *Head tube length* $(L_{HT_{bt}})$: This portion of the frame keeps the *Top Tube* and the *Down Tube* intact. When designing a track frame, the head tube ends up fairly long for a given frame size and is thus varied at two levels for experimentation at (95–100) for low level and (105–110) for high level.
5. *Head angle* (θ_h): Many track bikes go as steep at 72–76 (high level) degrees in head tube but a compromise can be made by decreasing it to 65–69 (low level) degrees for the road-cum-trek biking.
6. *Bottom bracket height* (L_{BB}): Most of the road bikes use 260–270 (low) *Bottom Bracket* height, whereas a similar make for track biking turns *Bottom Bracket* height at as high as 280–290 (high) from ground.
7. *Wheel radius* (r_w): Wheel radius was maintained at 285–300 for low level and 315–330 for high level.

9.6.2.2 Performance Comparison

Eight problem instances were generated with due considerations to the Taguchi's $L_8(2^7)$ OA as shown in Table 9.2 (Taguchi, 1987) by varying the aforementioned factors at two levels. The characteristic and search capability of TIRO is compared with that of GA, AGA, and SGA, and the results (from the test bed 1) are provided in Table 9.3 for detailed analysis. The success of the algorithm lies in the fact that the proposed metaheuristic adapts itself to select the best algorithm for each iteration whereas the pure algorithms are forced to continue with

Table 9.2 L_8 Orthogonal Array

Experiment No.	D_{ij}	ρ_m	C_{cul}	$L_{HT_{bt}}$	θ_h	L_{BB}	r_w
1	1	1	1	1	1	1	1
2	1	1	1	2	2	2	2
3	1	2	2	1	1	2	2
4	1	2	2	2	2	1	1
5	2	1	2	1	2	1	2
6	2	1	2	2	1	2	1
7	2	2	1	1	2	2	1
8	2	2	1	2	1	1	2

Table 9.3 Performance Comparison of GA, AGA, SGA, and Proposed TIRO

Experiment No.	GA	AGA	SGA	TIRO
1	290.322	298.963	302.571	315.17
2	350.712	363.503	362.994	371.604
3	259.926	271.831	268.797	281.39
4	283.54	293.121	289.113	299.796
5	322.683	329.871	340.473	344.738
6	459.768	455.412	460.198	472.934
7	450.663	453.481	466.718	509.782
8	451.889	473.529	461.967	492.806

themselves (as they have no choice) even if they may not perform better on the current population.

9.6.2.3 Analysis of Results

We divide the analysis in two sections: overall analysis utilizing ANOVA and detailed analysis of a particular problem instance.

9.6.2.3.1 ANOVA Analysis and F-Test

Table 9.4 lists the results obtained by five trials of the algorithms over each problem instance. On these values, ANOVA is performed and results are reported in Table 9.5. Moreover, F-test is also carried out at a relatively high significant level of 95%. It is observed from Table 9.4 that the value of F obtained from the model surpasses $F_{critical}$ (2.75), thus validating the robustness of the proposed TIRO on the underlying problem.

According to Phadke (1989), the ability of a factor to influence the objective function value depends upon its contribution to the total sum of squares, which has been used to determine the percentage contribution of the tested parameters. It is found that the percentage contribution of the demand for a product (D_{ij}) is momentously high to the objective function value. Similarly, other factors having some significant contribution to the objective function value are C_{cu}, $L_{HT_{bt}}$, and L_{BB}. Contribution of head angle (θ_h) was found to be very low; however, the choice of this factor is implicitly related to demand of a bicycle and is thus considered important for the study.

9.6.2.3.2 Analysis of Instance 1/Test Bed 1

To provide detailed insights and also to estimate convergence trends of the algorithms, the best fitness value obtained from TIRO, GA, AGA, and SGA are plotted against total number of fitness evaluations (Figure 9.5). It is evident from the same figure that TIRO significantly outperforms the remaining three techniques, thus validating its supremacy over other three approaches on the concerned problem.

Similarly, a detailed result obtained from the same problem instance (Instance 1/Test Bed 1) is provided in Table 9.6.

9.6.3 Test Bed 2

We generated test bed 2 in order to depict the robustness of the model to cope with nonuniformity in customer demand by establishing several demand modeling techniques. Ten problem instances were generated with due considerations to the nonuniformity of demand within the fixed market space. Demand in a static market space has been modeled as a discrete function, linear continuous function, normal

Table 9.4 Results (for Test Bed 1)

Instance Number	D_{ij}	ρ_m	C_{cul}	L_{HTbt}	θ_h	L_{BB}	r_w	Run 1	Run 2	Run 3	Run 4	Run 5	Average
1	1	1	1	1	1	1	1	312.55	316.24	320.81	310.06	316.19	315.17
2	1	1	1	2	2	2	2	370.35	371.58	367.62	375.98	372.49	371.604
3	1	2	2	1	1	2	2	280.75	278.56	281.67	282.34	283.63	281.39
4	1	2	2	2	2	1	1	300.41	298.16	295.98	301.49	302.94	299.796
5	2	1	2	1	2	1	2	340.94	343.82	350.12	347.44	341.37	344.738
6	2	1	2	2	1	2	1	472.19	468.79	475.32	478.14	470.23	472.934
7	2	2	1	1	2	2	1	510.57	512.46	507.26	513.28	505.34	509.782
8	2	2	1	2	1	1	2	492.66	488.35	495.64	493.26	494.12	492.806

Table 9.5 ANOVA Analysis (over Test Bed 1)

Parameter	Sum of Squares	d.f.	Mean Square	F	Percentage Contribution
D_{ij}	190,647.0563	1	190,647.0563	17,641.74	63.91051
p_m	3,933.08224	1	3,933.08224	363.9522	1.314937
C_{cul}	52,745.35876	1	52,745.35876	4,880.852	17.67918
L_{HTbt}	21,636.45225	1	21,636.45225	2,002.154	7.249964
θ_h	827.19025	1	827.19025	76.54499	0.273691
L_{BB}	20,976.4	1	20,976.4	1,941.075	7.028683
r_w	7,174.89796	1	7,174.89796	663.9374	2.40175
Model	298,286.2486	7	42,562.91967	3,938.608	—
Residual	345.81084	32	10.80658875	—	—
Total	298,286.2486	39	7,648.365	—	—

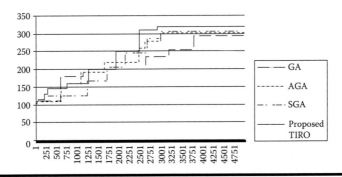

Figure 9.5 Plot of the best fitness value obtained from TIRO, GA, AGA, and SGA against total number of fitness evaluations.

distribution, and random function. Interested readers are referred to the PhD dissertation of Williams (2003) for detailed description of the demand scenarios.

9.6.3.1 Discrete Function

For discrete demand, the objective function is given by

$$O = \sum_{i}^{q} \frac{1}{D_i} \sum_{i=y_{i.\min}}^{y_{i.\max}} \sum_{j=x_{i.\min}}^{x_{i.\max}} \cdots \sum_{k=n_{i.\min}}^{n_{i.\max}} o\left(x_j, y_i, \ldots n_k\right) \tag{9.15}$$

Table 9.6 Best Results from Instance 1/Test Bed 1

Product/Variables	Mini	Average	Medium	Large
Stand-over height	26.15	28.75	30.6	32.3
Reach	25.45	26.6	27.45	28.15
θ_h	70.0	70.0	70.0	70.0
θ_s	72.5	72.5	72.5	72.5
r_w	12.80	13.78	13.78	14.76
l_{CS}	16.41	17.53	17.53	17.53
l_{SS}	15.63	15.45	16.38	15.73
l_{TT}	19.45	20.60	22.45	26.64
l_{DT}	24.02	25.64	26.68	29.88
l_{ST}	20.20	22.92	24.86	22.15
l_{HT}	4.17	4.17	4.17	4.17

where O denotes the objective function value, q denotes the total number of discrete steps (demand values) in the market space, and the upper and lower bounds of each discrete step is represented by the subscripts max and min.

9.6.3.2 Linear Continuous Function

For the case of linear demand, the objective function is given by

$$O = \sum_{i=y_{min}}^{y_{max}} \sum_{j=x_{min}}^{x_{max}} \cdots \sum_{k=n_{min}}^{n_{max}} \frac{1}{D\left(x_j, y_i, \ldots n_k\right)} o\left(x_j, y_i, \ldots n_k\right) \tag{9.16}$$

where $D(x, y)$ is in the form

$$D\left(x, y, \ldots n\right) = ax + by + \cdots + cn + d \tag{9.17}$$

and a, b, c, and d are coefficients used to make the approximations in the linear trends of demand.

9.6.3.3 N-Dimensional Gaussian Distribution

One-dimensional Gaussian distribution is given by following equation:

$$y = \frac{1}{\sigma\sqrt{2\pi}} e^{\frac{1}{2}\left(\frac{x-\mu}{\sigma}\right)^2} \tag{9.18}$$

where μ is the mean of the distribution at which maximum demand occurs and σ is the standard deviation of the distribution. Similarly, the demand for an n-dimensional market space is represented by the function given as

$$D(x,y) = \prod_{i}^{n}\left(A_i \frac{1}{\sigma_i\sqrt{2\pi}} e^{\frac{1}{2}\left(\frac{i-\mu_i}{\sigma_i}\right)^2}\right) \tag{9.19}$$

where subscript i denotes the statistical properties corresponding to each individual dimension and variable A represents a scaling coefficient used to translate the probability of receiving orders for a product to the demand of that product. The objective function for the present case is similar to Equation 9.16, with the demand function being replaced by Equation 9.19.

9.6.3.4 Random Distribution

For the case of random distribution, the objective function is similar to Equation 9.16. However, the demand of each individual product variant is unique and randomly distributed in a specified range.

In this research, we investigate 10 demand models derived from these 4 basic modeling techniques as discussed by Williams (2003). The demand for each of these 10 scenarios is taken as input to form a product platform. The best results obtained by the proposed search technique for each demand are reported in Table 9.7.

In essence, the aforementioned computational results not only statistically authenticate the efficacy and supremacy of the proposed TIRO search strategy but also provide a guideline to the manufacturer for dealing effectively with the nonuniform demand scenario.

9.7 Conclusion

In this chapter, we have considered the nonuniform model of the demand scenario for MTB frame fabrication problem that aims at maximizing the average profit of the manufacturer over the entire market space. An innovative search strategy (i.e., TIRO) was proposed and has been tested over two sets of problem instances

Table 9.7   Performance Comparison of GA, AGA, SGA, and Proposed TIRO (Test Bed 2)

Demand Scenario	GA	AGA	SGA	TIRO
Discrete pyramid large	292.2944	304.2755	307.2341	320.3369
Discrete pyramid small	278.5524	284.8841	289.4631	300.7033
Linear large	256.317	281.7969	285.9257	295.1037
Linear small	239.108	246.9709	246.7073	263.1777
Normal center small	290.322	298.963	302.571	315.17
Normal high	146.7258	156.1309	160.6521	169.191
Normal low	315.3053	334.8752	367.6783	409.1473
Normal center large	292.6274	320.4378	343.9172	360.5694
Random large	284.3041	303.6423	305.0805	313.5159
Random small	162.4772	167.3694	168.8724	174.8577

of varying size and complexity. The proposed TIRO search strategy provided consistently better results than its individual pure algorithm, thereby validating its superiority over the two approaches on the underlying problem. DOE techniques have been utilized to statistically study the obtained results as well as validate the algorithm's performance.

In essence, this research has twofold contributions to the existing platform literature and optimization technology, including the introduction of a case study of MTB frame design and the development of a TIRO framework for consistent generation of optimal/near-optimal results in the underlying platform planning problem. The roles of uncertainty within the market are not taken into account in the model formulation of the problem and are left as a scope for future study.

Appendix 9.A

Table 9.A.1   Notations

No.	Notation	Definition
1	θ_h	Head angle
2	θ_s	Seat angle

(continued)

Table 9.A.1 (continued)   Notations

No.	Notation	Definition
3	$L_{TT_{eff}}$	Effective top tube length
4	$L_{CS_{cc}}$	Chain stay CC
5	$L_{ST_{cc}}$	Seat tube CC
6	$L_{ST_{ct}}$	Seat tube CT
7	L_{BB}	Bottom bracket height
8	r_w	Wheel radius
9	$L_{HT_{cc}}$	Head tube length CC
10	$L_{HT_{bt}}$	Total head tube length
11	$L_{HT_{ctt}}$	Head tube above CTT
12	L_{FL}	Fork length
13	O_F	Fork offset CC
14	L_{LS}	Head set lower stack height
15	L_C	Crank length
16	W_{ra}	Rear axle width
17	W_{bb}	Bottom bracket shell width
18	d_{H_k}	Horizontal diameter for kth tube
19	d_{V_k}	Vertical diameter for kth tube
20	ρ_m	Density of material
21	h_{so}	Stand-over height
22	h_{cs}	Height of crank set above ground
23	l_R	Reach

Appendix 9.B

The mass of the MTB frame is calculated according to the frame geometry. Further, we define following seven coordinates for determining the length of each element of front triangle and rear triangle.

Table 9.B.1   Important Bike Coordinates

Point in 2D		Coordinate	Representation in Terms of Design Parameters
1	Bottom bracket	(x_1, y_1)	$(0, 0)$
2	Rear axle	(x_2, y_2)	$\left(\sqrt{L_{CS_{cc}}^2 - (r_w - L_{BB})^2}, \ (r_w - L_{BB}) \right)$
3	Front axle	(x_3, y_3)	$\left(L_{TT_{eff}} - (var_1/\tan(\theta_s)) + var_2, \ (r_w - L_{BB}) \right)$
4	Intersection of seat stay, seat tube, and top tube	(x_4, y_4)	$\left(-L_{ST_{cc}} \times \cos(\theta_s), \ L_{ST_{cc}} \times \sin(\theta_s) \right)$
5	Intersection of head tube and top tube	(x_5, y_5)	$\left(L_{TT_{eff}} - (var_1/\tan(\theta_s)), \ var_1 \right)$
6	Intersection of head tube and down tube	(x_6, y_6)	$\left(\begin{array}{l} L_{TT_{eff}} - (var_1/\tan(\theta_s)) + (L_{HT_{cc}} \times \cos(\theta_h)), \\ var_1 - (L_{HT_{cc}} \times \sin(\theta_h)) \end{array} \right)$
7	Effective seat tube top	(x_7, y_7)	$\left(-var_1/\tan(\theta_s), \ var_1 \right)$

$var_1 = \left(L_{FL} + L_{LS} - L_{HT_{bt}} - L_{HT_{ctt}} \right) \times \sin(\theta_h) - O_F \times \sin(\pi/2 - \theta_h) + L_{BB}.$

The lengths and thus the corresponding mass of *Chain stay, Seat stay, Top tube, Down tube, Seat tube,* and *Head tube* are given by $l_{CS}, l_{ST}, \ldots l_{HT}$, respectively.

$$l_{CS} = \left((x_2 - x_1)^2 + (y_2 - y_1)^2 \right) \tag{9.A.1}$$

$$l_{SS} = \left((x_2 - x_4)^2 + (y_2 - y_4)^2 \right) \tag{9.A.2}$$

$$l_{TT} = \left((x_6 - x_5)^2 + (y_6 - y_5)^2 \right) \tag{9.A.3}$$

$$l_{DT} = \left((x_6 - x_1)^2 + (y_6 - y_1)^2 \right) \tag{9.A.4}$$

$$l_{ST} = \left((x_4 - x_1)^2 + (y_4 - y_1)^2\right) = L_{ST_{cc}} \qquad (9.A.5)$$

$$l_{HT} = L_{HT_{bt}} \qquad (9.A.6)$$

The frame tubes are elliptical in shape with an independent horizontal (d_{h_k}) and vertical diameter (d_{v_k}) for each tube. Mass of the kth tube is given by the following formula:

$$W_{tube_k} = \rho_m \times \frac{\pi}{2} \times \left(d_{h_k} + d_{v_k}\right) t_{wall_k} \times l_{tube_k} \qquad (9.A.7)$$

where $k = 1$ stands for chain stay, $k = 2$ for seat stay, and so on, such that the total mass of the frame is given as

$$W_{metal} = \sum_{k=1}^{6} W_{tube_k} \qquad (9.A.8)$$

Appendix 9.C

Pseudocodes of GA, AGA, and SGA

Pseudocode of GA
Generate random population of solutions
For each individual: calculate **Fitness**
while(iter<iter_MAX){
    Perform **Crossover operation** based on probability of
    crossover;
    Perform **Mutation operation** based upon probability of
    mutation;
    Compute **Fitness**;
    Perform **Selection** operation for population of next
    generation.
    iter++;
}
Output: Best Solution of the problem

Pseudocode of AGA
Generate random population of solutions
For each individual: calculate **Fitness** and assign **age** = 0;
while(iter<iter_MAX){
    Perform **Crossover operation** based on probability of
    crossover;
    Perform **Mutation operation** based upon probability of
    mutation;

```
    Compute Fitness of each individual based on the age and
    objective function;
    Increment age by 1;
    Perform Selection operation for population of next
    generation.
    iter++;
}
Output: Best Solution of the problem

Pseudocode of SGA
Generate random population of solutions
For each individual: calculate Fitness
while(iter<iter_MAX){
    Select father from population by first selection scheme;
    Select mother from population by second selection scheme;
    Perform Crossover operation on father and mother;
    Perform Mutation operation based upon probability of
    mutation;
    Compute Fitness;
    Perform Selection operation for population of next
    generation.
    iter++;
}
Output: Best Solution of the problem
```

Authors

Mukul Tripathi is a graduate research assistant at Flexible Manufacturing and Lean Systems Laboratory at the University of Texas at San Antonio (UTSA). He received his bachelor's in technology degree from the National Institute of Foundry and Forge Technology, India, in 2008. His research interests include engineering management, workforce forecasting, artificial intelligence techniques and their variants, application of self-healing mechanism on manufacturing systems, modeling real-world disassembly sequencing problems, multiagent system design and application for multistation assembly processes, product platform planning and optimization, and reliability optimization. His mail id is <mukul.nifft@gmail.com>.

Hung-da Wan is an assistant professor of mechanical engineering and the director of Sustainable Manufacturing Systems Laboratory at the University of Texas at San Antonio (UTSA). He received his PhD in industrial and systems engineering from Virginia Tech. His research interests include lean manufacturing, computer-integrated manufacturing, and sustainability of manufacturing systems. He is among the core faculty of the Center for Advanced Manufacturing and Lean Systems (CAMLS) at UTSA. His mail id is <hungda.wan@utsa.edu>.

References

Allada, V., Choudhury, A.K., Pakala, P.K., Simpson, T.W., Scott, M.J., and Valliyappan, S. 2006. Product platform problem taxonomy: Classification and identification of benchmark problems. In: *2006 ASME Design Engineering Technical Conference*, Philadelphia, PA. ASME Paper No. DETC2006/DAC-99569.

Boyan, J. and Moore, A. 2000. Learning evaluation functions to improve optimization by local search. *Journal of Machine Learning Research* 1: 77–112.

Carchare, T. and Beck, J.C. 2005. Applying machine learning to low knowledge control of optimization algorithms. *Computational Intelligence* 21(4): 372–387.

Ghosh, A., Tsutsui, S., and Tanaka, H. 1996. Individual aging in genetic algorithms. *Australian and New Zealand Conference on Intelligent Information Systems*, Adelaide, SA, Australia.

Hernandez, G. 2001. Design of platforms for customizable products as a problem of access in a geometric space. PhD dissertation. George W. Woodruff School of Mechanical Engineering, Georgia Institute of Technology, Atlanta, GA.

Kumar, V.V., Tripathi, M., Pandey, M.K., and Tiwari, M.K. 2009a. Physical programming and conjoint analysis-based redundancy allocation in multistate systems: A Taguchi embedded algorithm selection and control (Tas&C) approach. *Proceedings of the Institution of Mechanical Engineers, Part O: Journal of Risk and Reliability* 223(3): 215–232.

Kumar, V.V., Tripathi, M., Tyagi, S.K., Shukla, S.K., and Tiwari, M.K. 2009b. An integrated real time optimization approach (IRTO) for physical programming based redundancy allocation problem. In: *3rd International Conference on Reliability and Safety Engineering*, Rajasthan, India, pp. 692–704.

Lagoudakis, M.G. and Littman, M.L. 2000. Algorithm selection using reinforcement learning. In: *Proceedings of 17th International Conference on Machine Learning*. Morgan Kaufmann, San Francisco, CA, pp. 511–518.

Mandal, S.K., Tyagi, S.K., Tripathi, M., and Tiwari, M.K. 2007. Optimization of series-parallel system reliability: Adaptive memetic particle swarm optimization based approach. In: *3rd International Conference on Reliability and Safety Engineering*, Rajasthan, India, pp. 681–691.

Mayer, M.H. and Lehnerd, A.P. 1997. *The Power of Product Platforms: Building Values and Cost Leaderships*. New York: Simon and Schuster Inc.

Nagalakshmi, M.R., Tripathi, M., Shukla, N., and Tiwari, M.K. 2009. Vehicle routing problem with stochastic demand (Vrpsd): Optimisation by neighbourhood search embedded adaptive ant algorithm (Ns-Aaa). *International Journal of Computer Aided Engineering and Technology* 1(3): 300–321.

Nanda, J., Thevenot, H.J., and Simpson, T.W. 2005. Product family representation and redesign: Increasing commonality using Formal Concept Analysis. In: *Proceedings of DETC'05, ASME 2005 International Design Engineering Technical Conferences and Computer and Information in Engineering Conference*, Long Beach, CA, September 24–28, 2005.

Phadke, M.S. 1989. *Quality Engineering using Robust Design*. Englewood Cliffs, NJ: Prentice Hall.

Pine, B.J. 1993. *Mass Customization: The New Frontier in Business Competition*. Boston, MA: Harvard Business School Press.

Rice, J. 1976. The algorithm selection problem. *Advances in Computers* 15: 65–118.

Ross, T.J. 1997. *Fuzzy Logic with Engineering Applications*, International Edition. New York: McGraw Hill.

Shukla, S.K., Tripathi, M., Kuriger, G., Wan, H., and Chen, F.F. 2009. Clonal C-fuzzy decision tree (C2FDT) for workforce deployment. *2009 Annual Industrial Engineering Research Conference*. May 30-June 3, Miami, FL, 2128–2133.

Simpson, T.W. 1998. A concept exploration method for product family design. PhD dissertation. Georgia Institute of Technology.

Simpson, T.W. 2005. Product platform design and customization: Status and promise. *Artificial Intelligence for Engineering Design, Analysis and Manufacturing*, 18(1): 3–20.

Taguchi, G. 1987. *System of Experimental Design 1&2*. New York: UNIPUB/Kraus International Publications.

Thevenot, H.J., Steva, E.D., Okudan, G.E., and Simpson, T.W. 2006. A multi-attribute utility theory-based approach to product line consolidation and selection. In: *Proceedings of IDETC/CIE 2006, ASME 2006, International Design Engineering Technical Conferences & Computers and Information in Engineering Conference*, Philadelphia, PA.

Tripathi, M., Agrawal, S., and Tiwari, M.K. 2008. Disassembly sequencing problem: Resolving the complexity by random search techniques. In: *Environment Conscious Manufacturing*, eds. S.M. Gupta and A.J.D. Lambert, pp. 331–362. Boca Raton, FL: CRC Press.

Tripathi, M., Agrawal, S., Pandey, M.K., Shankar, R., and Tiwari, M.K. 2009a. Real world disassembly modeling and sequencing problem: Optimization by algorithm of self-guided ants (Asga). *Robotics and Computer-Integrated Manufacturing* 25(3): 483–496.

Tripathi, M., Glenn, K., and Wan, H. 2009b. An ant based simulation optimization for vehicle routing problem with stochastic demands. In: *Winter Simulation Conference*, eds. M.D. Rossetti, R.R. Hill, B. Johansson, A. Dunkin, and R.G. Ingalls, Austin, TX.

Tsai, J.T., Liu, T.K., Chou, J.H., and Liu, T.K. 2004. Hybrid Taguchi-genetic algorithm for global numerical optimization. *IEEE Transactions on Evolutionary Computation* 8(4): 365–377.

Tsai, J.T., Liu, T.K., Chou, J.H., and Liu, T.K. 2006. Optimal design of digital IIR filters by using hybrid Taguchi genetic algorithm. *IEEE Transactions on Industrial Engineering* 53(3): 867–879.

Wagner, S. and Affenzeller, M. 2005. SexualGA: Gender-specific selection for genetic algorithms. In: *Proceedings of the 9th World Multi-Conference on Systemics, Cybernetics and Informatics (WMSCI)*, Orlando, FL, pp. 76–81.

Williams, C.B. 2003. Platform design for customizable products and processes with non-uniform demand. MS thesis. G.W. Woodruff Scholl of Mechanical Engineering, Georgia Institute of Technology, Atlanta, GA.

Williams, C.B., Allen, J.K., Rosen, D.W., and Mistree, F. 2004. Designing platforms for customizable products in markets with non-uniform demand. In: *Proceedings of the 2004 ASME Design Engineering Technical Conferences and Computers and Information in Engineering Conference*, Salt Lake City, UT, September 28–October 2, DETC2004/DTM-57469.

Chapter 10

Product Customization through the Development of Manufacturing Cores

Hazem Smadi, Ali K. Kamrani, and Sa'Ed M. Salhieh

Contents

This chapter discusses a model for manufacturing under the constraints and conditions of mass customization environment. The model is based on manufacturing features and involves the concept of modular design. That is, manufacturing features are identified and analyzed in a way that enables the generation of what is called "manufacturing core." Manufacturing cores are semifinished products that have certain manufacturing features. The core can be used to manufacture a range of products after conducting certain manufacturing processes. Manufacturing cores are generated through two phases of optimization. The first phase is known as product's manufacturing features analysis, which includes identification of starting features. The second phase is known as manufacturing core formation that ends with generation of manufacturing cores. The model is compared with make-to-order and make-to-stock policies.

10.1 Introduction

Mass customization is defined as manufacturing products based on individual customer needs while achieving the efficiency of mass production. New concepts of quality go beyond product performance and features to achieve customer satisfaction. A customer is satisfied if he can obtain a needed product at a reasonable price, while he is delighted if he can obtain a needed product that is customized to his individual requirements (Silveira et al., 2001; Tseng and Jiao, 2004). Customized products are manufactured in small batches and based on the request from the customer. The cost of customized products is higher than that for standard ones because manufacturing customized products entails certain design and manufacturing activities in order to meet individual needs (Gu et al., 2002). Standard products enjoy the benefits of economy of scale by production of standard components in large quantities to meet several individual needs. Manufacturing customized products in large quantities to meet individual customer needs is a challenge to industry. Mass customization achieves customer satisfaction, but it needs certain enabling and control tools (Radder and Louw, 1999; Berman, 2002).

Global competition has necessitated introducing the concepts of high variety products with the minimum time to market with a high variety of customer needs. Organizations are faced with the challenge of new technology development and constraints that do not permit them to be strong competitors (Jiao et al., 2003). Competitive organizations introduce a wide variety of products to meet the diversity of customer needs. Variant product manufacturing means higher costs than standard products. Mass production system is considered to be a highly efficient system as production is undertaken in the economy of scale, which is attained by producing high quantities of standard products (Radder and Louw, 1999). Mass customization helps organizations responding rapidly to individual customer requirements. The problem faced in this system is to produce high customized products while achieving the efficiency of mass production (Yang and Li, 2002). The mass customization system includes three major areas within organizations: design, manufacturing, and sales and distribution. The design area includes customer demand identification and analysis as well as the design issues under the constraints of the customer. The sales and distribution area includes all the activities related to customer–organization relationship after product manufacture, such as delivery and after-sale services. The manufacturing area is somehow complicated and can be described by process planning and management activities that are needed to manufacture a customized product, leading to a custom route and higher lead time (Tseng and Jiao, 2001).

This chapter presents an initial work done on a model for manufacturing under the constraints and conditions of mass customization environment. The proposed model is based on manufacturing features and involves the concept of modular design. That is, manufacturing features are identified and analyzed in a way that enables the generation of what is called "manufacturing core." Manufacturing cores are semifinished products that have certain manufacturing features. The core can be used to manufacture a range of products by performing certain manufacturing processes. Manufacturing cores are generated through two phases of optimization. The first phase is known as product's manufacturing features analysis, which includes identification of starting features. The second phase is known as manufacturing core formation, which ends with the generation of manufacturing cores.

10.2 Methodology

The methodology consists of two phases aimed at determining the best set of manufacturing features to be included in manufacturing cores in order to be used in the manufacturing of all products. These cores are to be stocked as semifinished products. Phase one aims at identifying a set of manufacturing features known as starting features. Phase two aims at identifying the optimal core sets. Figure 10.1 illustrates the proposed methodology.

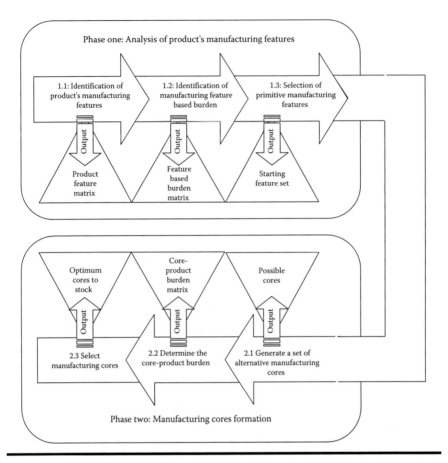

Figure 10.1 Methodology phases.

10.2.1 Phase One: Analysis of Product's Manufacturing Features

Phase one starts with identifying the manufacturing features for each individual product. Thus, the feature-based burden required to define a feature from a solid workpiece or another feature can be determined. The phase ends with identifying a set of primitive manufacturing features known as starting features. This set of features is used as an input for phase two.

10.2.1.1 Identification of Product's Manufacturing Features

Product features are distinct geometrical descriptions or shapes that can give a complete topographical representation of products. Identifying product features mainly focuses on identifying a basic set of entities that could be combined to form a product. Entities could be a set of solid primitives like those used to represent solid models using the constructive solid geometry. Manufacturing features are identified by decomposing a product into a set of standard entities that are attainable by some existing manufacturing operation. For example, a hole is a manufacturing feature that is achievable by a drilling operation. The distinction between design and manufacturing features is sometimes hard to make and can be somewhat confusing. But one can say that a design feature is used to build the product irrespective of how it will be manufactured, while a manufacturing feature is based on feasible manufacturing operations and thus provide a direct link to manufacturing processes. Identifying manufacturing features usually begins by investigating the geometrical description of the operations and then a set of manufacturing operations capable of shaping the products starting from a raw workpiece into its intended geometry are determined (Li et al., 2002). After that, the manufacturing operations are grouped based on their similarities and a minimum set of manufacturing operations are generated. Finally, a basic set of generic manufacturing features are identified and used to describe the product. Each type of the generic manufacturing features can have a set of variants or specific features based on some dimensions. The generic manufacturing features can be thought of as high-level features, while the variant features are grouped as low-level features specifying the final dimensions of the manufacturing features. The generic features and their respective specific variants are used to describe the products under study using a matrix as shown in Figure 10.2, where 1 indicates that the manufacturing feature is used to describe the product, and a blank cell indicates that it is not.

10.2.1.2 Identification of Manufacturing Feature-Based Burden

Manufacturing process involves efforts to convert raw materials to a finished product. These efforts include the time needed for manufacturing and the cost of manufacturing. The manufacturing time encompasses the setup time, processing time, and handling time. The manufacturing cost includes the cost of tools, labor, and consumed materials. The total effort for manufacturing a product includes the time needed for manufacturing and the cost of manufacturing, which is known as the feature-based burden of manufacturing. Manufacturing feature-based burden is defined as the total effort needed to manufacture a product. The manufacturing feature-based burden can be considered as a function of the manufacturing efforts. This function includes the components of manufacturing efforts in different weights depending on the nature of the manufacturing process. The details of calculating the values of each effort components and setting the relative weights are beyond the scope of this chapter and will not be addressed. The objective of this step is to identify the manufacturing feature-based burden for each feature from a solid workpiece or from other features. Identifying feature-based burden is important to determine the feasible set of starting features because it may be feasible to

Generic features	Variant features	P1	P2	P3	...
G1	F1	1		1	...
	F2	1			...
	F3	1	1	1	...
G2	F4		1		...
	F5	1		1	...
	F6	1	1		...
G3	F7		1		...
	F8	1		1	...
G4	F9	1		1	...
:	:	:	:	:	...
:	M	1			...

P: Product
G: Generic feature
F: Variant feature

Figure 10.2   The product-feature matrix.

manufacture a feature from more than one feature, so the decision will rely on the feature-based burden as an input for the optimization in the next step. All manufacturing features can be manufactured from a solid raw workpiece, but this may lead to a very high manufacturing cost, especially if machining was the main manufacturing operation used. So if machining was to be used, it is better to select the raw workpiece such that it contains some preexisting features that are developed from a bulky process like casting or forging. Using these workpieces (with preexisting features) will significantly reduce the manufacturing feature-based burden. The same idea can be extended to address the feasibility of manufacturing product features from some premanufactured features. For example, a large hole could be made by enlarging a smaller hole, this enlargement of the hole will be at a certain feature-based burden (cost) that is expected to be less than the feature-based burden of making the large hole from a solid workpiece. On the other hand, a small hole cannot be made from a large hole. The idea of estimating the feature-based burden of making a product's features from existing features is better illustrated in

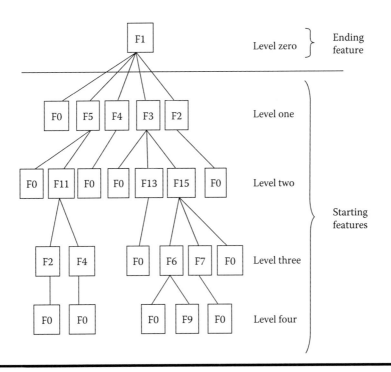

Figure 10.3 Hierarchy of product features.

Figure 10.3 where a hierarchy of product features is arranged in four levels as an example. Level zero has the product's final (or ending) features. Level one has the starting features of the ending feature F1. That is F1 can be made from F0, F5, F4, F3, or F2. Level two shows the starting features of the features in level one. This implies that lower levels in the hierarchy show the starting features of higher levels. It can be observed from Figure 10.3 that each feature can be manufactured from F0, which represents a solid workpiece. The feasible features to manufacture F3 are F15 and F13. Also feature F15 can be manufactured from F7 and F6. Feature F6 can be manufactured from F9.

The features in the feature hierarchy can be represented using two vectors. The starting vector F_s contains the set of all starting features for all features:

$$F_s = \{F_{s1}, F_{s2}, F_{s3}, \ldots, F_{sn}\} \tag{10.1}$$

The second vector is the ending features vector F_e, which represents all the features that are incorporated in a product, made from a starting feature or from a solid workpiece:

$$F_e = \{F_{e1}, F_{e2}, F_{e3}, \ldots, F_{em}\} \tag{10.2}$$

Feature-based burden is calculated for each possible path. Feature-based burden is denoted by B_{ij}, which represents the feature-based burden to make feature j from feature i, when it is feasible. For feature-based burden calculations and analysis, level one in the starting features in Figure 10.3 is observed to identify the feasible features to make another feature. From Figure 10.3, F1 can be manufactured from F0, F2, F3, F4, and F5. This means that it is not feasible to make F1 from F13. It should be noted here that F13 can be used to make F3, and F3 can be used to make F1. Thus, it is possible to make F1 from F13 if and only if F3 was made as an intermediate feature to link F1 with F13. Feature-based burden calculations depend on the path selected and not on the number of features included in the path. For example in Figure 10.3, F1 can be made from F4 and F5, but F5 follows the path of F11 and F4, which means that the path to make F1 consequently is F4 then F11 then F5 then F1. There is no relationship between the feature-based burden to make F1 directly from F4 or to make F1 from F5, F11, and F4. The feature-based burden of the first path may be less, or greater than the feature-based burden of the second path. This implies that the feature-based burden from F4 to F1 is not equal to the addition of the feature-based burden following F4, F11, F5, then F1. The feature-based burden links the elements of the starting features and the ending features. The feature-based burden is represented by a matrix, which is generated by combining the starting features vector F_s and the ending features vector F_e. The generated matrix is shown in Figure 10.4, that is, the feature-based burden of making ending features from starting features. For example, the feature-based burden of making feature F3 from feature F2 is B_{23}, while it is not feasible to make feature F1 from feature F3, so the corresponding cell in the matrix denoted by M, which represents infeasibility.

Fe \ Fs	1	2	3	......	....	......	......	M
0	B_{01}	B_{02}	B_{03}	......	....	......	......	B_{0m}
1	B_{11}	B_{12}	M	......	....	......	......	B_{1m}
2	B_{21}	M	B_{23}	......	....	......	......	B_{2m}
:	......	......	......	......	....	......	......	......
:	......	......	......	......	....	......	......	......
:	......	......	......	......	....	......	......	......
:	......	......	......	......	....	......	......	......
n	B_{n1}	B_{n2}	B_{n3}	......	....	......	......	B_{nm}

Figure 10.4 Feature-based burden matrix.

10.2.1.3 Selection of Primitive Manufacturing Features

The objective of this step is to identify the optimum starting features from the whole set of starting features to make all ending features. The optimization is conducted using a linear programming model. The objective of the model is to minimize the summation of the feature-based burdens that are feasible to make all ending features.

Objective function:

$$\text{Min} \sum_{i=0}^{n} \sum_{j=1}^{m} B_{ij} * X_{ij} \qquad (10.3)$$

where

$$X_{ij} = \begin{bmatrix} 1, & \text{if feature j is made from feature i} \\ 0, & \text{otherwise} \end{bmatrix}$$

B_{ij} = the feature-based burden to make a feature j from feature i

Constraint 1: All features in a product are manufactured.

$$\sum_{i=0}^{h} \sum_{j=1}^{z} X_{ij} \geq \text{no. of features per product}, \quad \forall \text{ products} \qquad (10.4)$$

where

 i represents possible starting features for ending features in each product
 j represents ending features in each product
 h represents the limit of starting features in each product
 z represents the limit of ending features in each product

Constraint 1 guarantees the manufacturing of all products without a loss in any feature. That is, an ending feature should be made at least from one starting feature.

Constraint 2: Complete product–feature connectivity within a product for all products.

$$\sum_{i=0}^{n} X_{ij} (\text{for all j}) - \sum_{i=0}^{n} X_{ij} (\text{for all j}) = 0, \quad \forall \text{ products} \qquad (10.5)$$

This constraint ensures that the correct number of ending features is maintained within and across all products. The benefit of this constraint can be explained by Figure 10.7.

Constraint 3: X_{ij} is binary: 1 if it exists, 0 otherwise.

$$X_{ij} = (0, 1) \tag{10.6}$$

The output of phase one identifies a set of starting features.

10.2.2 Phase Two: Manufacturing Cores Formation

Phase two aims at selecting an optimum set of cores from the alternative set of cores. The selection is conducted through three steps. The starting features that resulted from phase one are used as input in determining the alternative cores. Then, the product-based burden of each core is calculated in order to optimize for alternative cores. The output of phase two is a set of cores. These cores should incorporate certain features that enable manufacturing the set of all products after implementing certain manufacturing processes.

10.2.2.1 Generate a Set of Alternative Manufacturing Cores

A manufacturing core is a semifinished product having a set of starting features that can be further manufactured to generate a range of different products. The needed manufacturing processes depend on the features of the cores and the features of the finished product. The importance of a core comes from manufacturing and stocking a semifinished product that will cover a range of products. It is not needed to stock the finished products. Stocking semifinished products differentiates a range of products. For inaccurate demand, considering more demand than forecasted for a product and less demand than forecasted for another product, the same core (semifinished product) can be used to balance the demand against forecast because the core is used for the product that has a higher demand. Also, stocking semifinished products needs less manufacturing time to make a finished product than when following a make-to-order policy. Several manufacturing cores could be found for a group of products by using a complete enumeration of all feasible paths to make each feature. Feasible paths are the set of the all starting features that can be used to make an ending one. Alternative cores can be generated by the combination of the starting features generated by phase one, but sometimes (depending on the case) the number of starting features could be relatively high, which leads to a very large number of combinations and a complex problem. Figure 10.5 shows a product-feature structure to illustrate the feasible paths to manufacture a feature in a product. The product in Figure 10.5 contains F7, F12, and F16 as features, which are known as ending features. The starting features are considered to be included in the core module that enables manufacturing the product from the core module. There is more than one feasible core module including starting features to manufacture the product. The feasible core modules are generated by alternating the starting features for each ending feature each time. The feasible paths to make feature F7 in product P are from F0 and F2; to make F12 are from F0, F6, and F8; and to

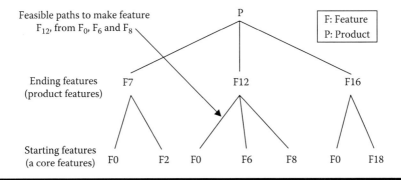

Figure 10.5 Product-feature structure.

make F16 are from F0 and F18. A feasible core module contains F2, F6, and F18 as starting features, which enables making F7, F12, and F16, respectively. Also, it is not feasible to include variant features that are at the same position. For example, consider two holes with the same position but different diameters. It is infeasible to have a core module that contains those two variant features. The list of possible cores for the product P in Figure 10.5 is shown in Table 10.1.

Table 10.1 List of Possible Core for the Product P in Figure 10.5

Possible Cores	Ending Features			
	F_7	F_{12}	F_{16}	
Y_1	F_0	F_0	F_0	Starting features for each ending feature in the product (core features)
Y_2	F_2	F_0	F_0	
Y_3	F_0	F_6	F_0	
Y_4	F_0	F_8	F_0	
Y_5	F_0	F_0	F_{18}	
Y_6	F_0	F_6	F_{18}	
Y_7	F_0	F_8	F_{18}	
Y_8	F_2	F_6	F_0	
Y_9	F_2	F_8	F_0	
Y_{10}	F_2	F_0	F_{18}	
Y_{11}	F_2	F_6	F_{18}	
Y_{12}	F_2	F_8	F_{18}	

10.2.2.2 Determine the Core Product–Based Burden

A manufacturing core can be differentiated from other products by applying the needed manufacturing processes capable of converting the manufacturing features available in the core into manufacturing features required in the final products. This process entails certain burden, which can be estimated by calculating the amount of burden resulting from manufacturing a complete product from a certain manufacturing core. This burden is called the core product–based burden. The core product–based burden of manufacturing products can be represented in a matrix as shown in Figure 10.6, where Y_k denotes for a possible cores where k is the index starting from 1 to r; C_{Pk} denotes for the product-based burden to make products (P) from a core (k); and C_k is the total product-based burden for making all products from a core (k).

Core product–based burden matrix is considered as the input for phase two. If it is not feasible to manufacture a product (P) from Y_k, the corresponding cell in the matrix is replaced by a very big number (very large product-based burden) denoted by M.

For example, consider products P1, P2, and P3 having the following features:

P1 having (F1, F2, F4)
P2 having (F2, F3, F4)
P3 having (F3, F4, F6)

And the alternative cores are:

Y1 having (F1, F2)
Y2 having (F3, F4)

Based on Figure 10.10, to manufacture P1, it is needed only to make F4 on core Y1, then the value of C_{11} in the core product–based burden matrix is the cost of making F4 from a solid workpiece, while it is infeasible to make P1 from Y2. To

	Y1	Y2	Y3	....	....	....	Y_r	
P1	C_{11}	C_{12}	M	....	....	....	C_{1r}	Y: Core
P2	M	C_{22}	C_{23}	....	....	....	C_{2r}	P: Product
P3	M	M	C_{33}	....	....	....	C_{3r}	B: Burden
Total cost	C_1	C_2	C_3	....	....	....	C_r	

Figure 10.6 Core product–based burden matrix.

manufacture P2, it is needed only to make F2 on core Y2, then the value of C_{22} in the core product–based burden matrix is the cost of making F2 from a solid workpiece, while it is infeasible to make P2 from Y1.

10.2.2.3 Select Manufacturing Cores

The objective of this step is to select the minimum set of cores from the set of alternative cores that can be used to manufacture the products. The selection of the best set of cores from a large number of possible cores resembles the problem solved by the "set covering algorithm," in which the activities are represented by the cores and the characteristics are represented by the products, and the objective is to optimize for covering products by the alternative core. The set covering algorithm is used to select the cores to manufacture all products. The objective is to select the minimum number of cores that have the least manufacturing product-based burden. The core product–based burden is denoted by C_k, which is the summation of the product-based burdens related for product for each core.

Objective function: To minimize the total product-based burden of manufacturing all products form each alternative core.

$$\text{Min} \sum_{k=1}^{r} C_k * Y_k \tag{10.7}$$

where

$$Y_k = \begin{bmatrix} 1, & \text{if the core k is used} \\ 0, & \text{otherwise} \end{bmatrix}$$

C_k is the total product-based burden for making all products from a core (k)
(k) is from 1 to r

Constraint 1: Each product is manufactured from at least one core:

$$\sum_{k=1}^{r} Y_k \geq 1 \quad \text{for all P} \tag{10.8}$$

This constraint ensures that a product will be manufactured from at least one core.

Constraint 2: Minimum number of cores to be used.

$$\sum_{k=1}^{r} Y_k \leq N_p \tag{10.9}$$

where N_p is the number of cores that are going to be used to manufacture all products.

This constraint specifies the maximum number of cores to be selected. N_p will be manually varied until reaching the minimum value of the objective function. Determining the optimal value of N_p is beyond the scope of this chapter.

Constraint 3: Y_k, is a binary variable, 1 if it exists, 0 otherwise

$$Y_k = (0,1) \tag{10.10}$$

The output of phase two is (are) the core(s) that will be stocked and used to manufacture a range of products. The next section presents a case study for the previous methodology showing the importance of manufacturing cores.

10.3 Case Study

The developed model was tested using a group of products to verify the functionality of the model. The selected group of products is flanges because they are simple and contain enough features that can address the functionality of the model. Generic and variant features exist in the selected flange products. Generic features are of three types: hole, step, and base. There are 11 variant features of the hole, 8 variant features of the step, and 5 variant features of the base. This forms a total of 24 variant features. The products of the case study are shown in Figure 10.7. The steps of the methodology applied on the flange products are as shown in the following sections.

10.3.1 Phase One: Analysis of Product's Manufacturing Features

10.3.1.1 Identification of Product's Manufacturing Features

The product–feature matrix is identified and shown in Figure 10.8. The details of each product's features regarding the dimension, position, and other specifications are shown in Figure 10.9.

10.3.1.2 Identification of Manufacturing Feature–Based Burden

The starting and ending features are identified by determining the feasibility to make each individual feature from another individual feature. The starting features vector is

$$Fs = \{F_0, F_1, F_7, F_{12}, F_{13}, F_{14}, F_{18}, F_{20}, F_{22}, F_{23}\} \tag{10.11}$$

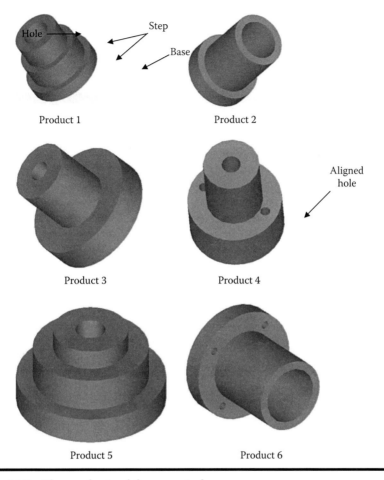

Figure 10.7   The products of the case study.

The ending features vector:

$$Fe = \{F_1, F_2, F_3, F_4, F_5, F_6, F_7, F_8, F_9, F_{10}, F_{11}, F_{12}, F_{13}, F_{14}, F_{15}, F_{16},$$

$$F_{17}, F_{18}, F_{19}, F_{20}, F_{21}, F_{22}, F_{23}, F_{24}\} \tag{10.12}$$

For the feasible feature paths, the feature-based burden to make a feature from another individual feature is calculated to generate the feature-based burden matrix. For calculation simplicity, it is assumed that working is on a workpiece:

1. To make a hole, every 1 mm needs
   0.0104 h to make the diameter
   0.00936 h to make the length

		P1	P2	P3	P4	P5	P6
Hole features	F1	1	1	0	0	0	0
	F2	0	0	1	0	0	0
	F3	0	0	0	1	0	0
	F4	0	0	0	1	0	0
	F5	0	0	0	1	0	0
	F6	0	0	0	0	1	0
	F7	0	0	0	0	0	1
	F8	0	0	0	0	0	1
	F9	0	0	0	0	0	1
	F10	0	0	0	0	0	1
	F11	0	0	0	0	0	1
Step features	F12	1	0	0	0	0	0
	F13	1	0	0	0	0	0
	F14	0	1	0	0	0	0
	F15	0	0	1	0	0	0
	F16	0	0	0	1	0	0
	F17	0	0	0	0	1	0
	F18	0	0	0	0	1	0
	F19	0	0	0	0	0	1
Base features	F20	1	0	0	0	1	0
	F21	0	1	0	0	0	0
	F22	0	0	1	0	0	0
	F23	0	0	0	1	0	0
	F24	0	0	0	0	0	1

Figure 10.8 Product-feature matrix for flanges.

2. When working at the circumference of a workpiece, every 1 mm needs
0.0078 h when working for the diameter
0.00702 h when working at the length
3. It is assumed that the solid workpiece, which holds the index F_0, is a cylindrical raw piece with a diameter of 125 mm and a length of 150 mm.
4. Then the feature-based burden is calculated based on a rate of \$15/mm removed.

Figure 10.10 shows the feature-based burden matrix. As shown in Figure 10.10, the feature-based burden to make feature F1 from feature F0 (solid workpiece) is \$17.35. Shown below are the calculations for this value:

F0 (d = 125 mm, L = 150 mm).
F1 (d = 40 mm, L = 95 mm), (0,0) position in (X,Y) axis with respect to the center of the cylindrical raw piece.
It needs (0.0104 × 40 = 0.416) h to make the diameter of F1.

Hole			Diameter	Length	Position	
					X	Y
	P1	H1	40	95	0	0
	P2	H2	40	95	0	0
	P3	H3	15	70	0	0
	P4	H4	15	85	0	0
		H5	8	35	27	0
		H6	8	35	−27	0
	P5	H7	22	60	0	0
	P6	H8	35	65	0	0
		H9	5	13	28	0
		H10	5	13	−28	0
		H11	5	13	0	28
		H12	5	13	0	−28
Step	P1	S1	70	35		
		S2	90	35		
	P2	S3	55	65		
	P3	S4	45	45		
	P4	S5	45	50		
	P5	S6	55	15		
		S7	85	25		
	P6	S8	45	51		
Base	P1	B1	110	20		
	P2	B2	70	20		
	P3	B3	85	23		
	P4	B4	75	35		
	P5	B5	110	20		
	P6	B6	70	13		

Figure 10.9 The details of each product's features regarding the dimension, position, and other specifications.

It needs $(0.0078 \times 95 = 0.741)$ h to make the length of F1.

The total time needed equals $(0.416 + 0.741 = 1.157\,\text{h})$, this leads to a feature-based burden of $(1.157\,\text{h} \times \$15 = \$17.355)$.

10.3.1.3 Selection of Primitive Manufacturing Features

The aim of the objective function is to minimize the total feature-based burden to make all ending features from all feasible starting features for all products (Smadi, 2005). The model was solved using LINDO 6.1* optimization package. The output of phase one identifies a set of starting features which are

---

* Lindo 6.1 is a registered trademark of LINDO Systems, Inc.

	F1	F2	F3	F4	F5	F6	F7	F8	F9	F10	F11	F12	F13	F14	F15	F16	F17	F18	F19	F20	F21	F22	F23	F24
F0	17.35	10.53	12.28	5.34	5.343	10.45	13.65	2.31	2.31	2.31	2.31	19.83	17.23	18.77	22.28	21.76	24.43	18.77	21.76	15.79	21.41	18.98	19.12	22.14
F1	M	6.14	4.56	M	M	6.21	3.06	M	M	M	M	M	M	M	M	M	M	M	M	M	M	M	M	M
F2	M	M	M	M	M	M	M	M	M	M	M	M	M	M	M	M	M	M	M	M	M	M	M	M
F3	M	M	M	M	M	M	M	M	M	M	M	M	M	M	M	M	M	M	M	M	M	M	M	M
F4	M	M	M	M	M	M	M	M	M	M	M	M	M	M	M	M	M	M	M	M	M	M	M	M
F5	M	M	M	M	M	M	M	M	M	M	M	M	M	M	M	M	M	M	M	M	M	M	M	M
F6	M	M	M	M	M	2.35	M	M	M	M	M	M	M	M	M	M	M	M	M	M	M	M	M	M
F7	M	M	M	M	M	M	M	M	M	M	M	M	M	M	M	M	M	M	M	M	M	M	M	M
F8	M	M	M	M	M	M	M	M	M	M	M	M	M	M	M	M	M	M	M	M	M	M	M	M
F9	M	M	M	M	M	M	M	M	M	M	M	M	M	M	M	M	M	M	M	M	M	M	M	M
F10	M	M	M	M	M	M	M	M	M	M	M	M	M	M	M	M	M	M	M	M	M	M	M	M
F11	M	M	M	M	M	M	M	M	M	M	M	M	M	M	M	M	M	M	M	M	M	M	M	M
F12	M	M	M	M	M	M	M	M	M	M	M	M	M	M	M	M	4.21	1.75	M	M	M	M	M	M
F13	M	M	M	M	M	M	M	M	M	M	M	M	M	M	3.51	2.98	7.02	M	2.98	M	M	M	M	M
F14	M	M	M	M	M	M	M	M	M	M	M	M	M	M	M	M	M	M	M	M	M	M	M	M
F15	M	M	M	M	M	M	M	M	M	M	M	M	M	M	M	M	M	M	M	M	M	M	M	M
F16	M	M	M	M	M	M	M	M	M	M	M	M	M	M	M	M	M	M	M	M	M	M	M	M
F17	M	M	M	M	M	M	M	M	M	M	M	M	M	M	M	M	5.26	M	M	M	M	M	M	M
F18	M	M	M	M	M	M	M	M	M	M	M	M	M	M	M	M	M	M	M	M	M	M	M	M
F19	M	M	M	M	M	M	M	M	M	M	M	M	M	M	M	M	M	M	M	M	M	M	M	M
F20	M	M	M	M	M	M	M	M	M	M	M	M	M	M	M	M	M	M	M	M	M	M	M	M
F21	M	M	M	M	M	M	M	M	M	M	M	M	M	M	M	M	M	M	M	M	M	M	M	63.53
F22	M	M	M	M	M	M	M	M	M	M	M	M	M	M	M	M	M	M	M	M	2.42	M	M	31.59
F23	M	M	M	M	M	M	M	M	M	M	M	M	M	M	M	M	M	M	M	M	2.28	M	M	3.01
F24	M	M	M	M	M	M	M	M	M	M	M	M	M	M	M	M	M	M	M	M	M	M	M	M

Figure 10.10 The feature-based burden matrix.

$$Fs = \{F_0, F_1, F_7, F_{12}, F_{13}, F_{14}, F_{23}\} \tag{10.13}$$

10.3.2 Phase Two: Manufacturing Cores Formation

10.3.2.1 Generate a Set of Alternative Manufacturing Cores

The structure in Figure 10.11 is used to generate the set of alternative manufacturing cores that are related to product 1. The same approach is used to generate the other manufacturing cores.

Each individual product can be manufactured from one feature or more of the feature groups and the combinations of individual features shown in Table 10.2.

Table 10.3 shows the features of the possible cores and the products that can be manufactured from each core. The possible cores are generated by combining the features for each product shown in Table 10.2.

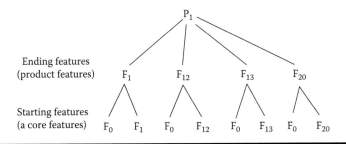

Figure 10.11   Product-feature structure for product 1.

Table 10.2   Features That Can Be Used to Manufacture Each Product

Product	The Features That Can Be Used to Manufacture the Product
P1	F_1, F_{12}, F_{13}
P2	F_1, F_{14}, F_{23}
P3	F_1, F_7, F_{14}
P4	F_1, F_{14}, F_{23}
P5	F_1, F_7, F_{12}, F_{13}
P6	F_1, F_7, F_{14}, F_{23}

10.3.2.2 Determine the Core Product–Based Burden

For simplicity, the core product–based burden is represented by the effort needed to manufacture a product from a core based on the time needed. Each entry in the core product–based burden matrix is calculated based on the needed manufacturing process to be conducted on a core in order to manufacture certain product. Following is a sample of the calculations of the entries in the core cost matrix:

Table 10.3   Possible Cores Generated

Possible Core	Features in the Core	Products That Can Be Manufactured by the Core
Y_1	F_0	P1, P2, P3, P4, P5, P6
Y_2	F_1	P1, P2, P3, P4, P5, P6
Y_3	F_7	P5, P6
Y_4	F_{12}	P1, P5
Y_5	F_{13}	P1, P5
Y_6	F_{14}	P2, P3, P4, P6
Y_7	F_{23}	P2, P4, P6
Y_8	F_1, F_{12}	P1, P5
Y_9	F_1, F_{13}	P1, P5
Y_{10}	F_1, F_{14}	P2, P3, P4, P6
Y_{11}	F_1, F_{23}	P2, P4, P6
Y_{12}	F_7, F_{12}	P5
Y_{13}	F_7, F_{13}	P5
Y_{14}	F_7, F_{23}	P6
Y_{15}	F_7, F_{14}	P6
Y_{16}	F_{14}, F_{23}	P2, P4, P6
Y_{17}	F_{12}, F_{13}	P1, P5
Y_{18}	F_1, F_{12}, F_{13}	P1, P5
Y_{19}	F_1, F_{14}, F_{23}	P2, P4, P6
Y_{20}	F_7, F_{12}, F_{13}	P5
Y_{21}	F_7, F_{14}, F_{23}	P6

The core Y_{10} contains (F_1, F_{14}):

F_1: a centered hole of diameter 40 mm and length of 95 mm
F_{14}: a step of diameter 55 mm and length of 65 mm

The features of P2 are as follows:

F_1: a centered hole of diameter 40 mm and length of 95 mm
F_{14}: a step of diameter 55 mm and length of 65 mm
F_{21}: a base of diameter 70 mm and length of 20 mm

This means that it is needed only to make the base (of diameter 70 mm and length of 20 mm) on Y_{10} to make P2.

Considering working the base from a solid model, the cost to make the base is the cost to make F_{21} from F_0, which equals ($21.41). Figure 10.12 shows the product core burden matrix.

10.3.2.3 Select Manufacturing Cores

The aim of the objective function is to minimize the total product-based burden of manufacturing all products from each alternative core (Smadi, 2005). The solution of phase two is to stock Y10 and Y18 as cores to manufacture P2, P4, P6, and P1, P5, respectively. The value of N_p was varied to select the number of cores needed. The cost varies for each product based on the number of cores used. Figure 10.13 shows a comparison when changing the number of cores. The optimal number of cores is when N_p equals two, as shown by the total cost in Figure 10.13. It should be noted here that one core can be used to manufacture all products but the cost is higher than that when using two cores. Having more than two cores provides the advantage of more flexibility through manufacturing a product from more than one core. For example, product 5 can be manufactured from core number 13 and core number 18 when using four cores, but the total costs are more than when compared to using two cores.

10.3.3 *Manufacturing Policies Comparison*

This section compares the proposed model of mass customization regarding manufacturing with make-to-stock and make-to-order manufacturing policies. Make-to-stock and make-to-order policies need special characteristics. Make-to-stock covers the demand with zero lead time but involves the cost of holding items in stores depending on the holding period and the item cost. But, in make-to-order policy, there is a lead time but there is no holding cost because there are no stocked items. The comparison relies on two different scenarios:

	Y1	Y2	Y3	Y4	Y5	Y6	Y7	Y8	Y9	Y10	Y11	Y12	Y13	Y14	Y15	Y16	Y17	Y18	Y19	Y20	Y21
																		F13	F23	F13	F23
								F12	F13	F14	F23	F12	F13	F23	F14	F23	F13	F12	F14	F12	F14
0		F1	F7	F12	F13	F14	F23	F1	F1	F1	F1	F7	F7	F7	F7	F14	F12	F1	F1	F7	F7
P1	70.22	52.86	M	50.39	52.98	M	M	33.03	35.63	M	M	M	M	M	M	M	33.15	15.80	M	M	M
P2	57.54	40.19	M	M	M	38.77	38.42	M	M	21.41	21.06	M	M	M	M	19.64	M	M	2.28	M	M
P3	51.81	47.42	M	M	M	33.03	M	M	M	28.64	M	M	M	M	M	M	M	M	M	M	M
P4	63.86	56.14	M	M	M	45.08	44.73	M	M	37.36	37.01	M	M	M	M	25.95	M	M	18.23	M	M
P5	69.46	65.22	61.36	49.24	52.05	M	M	45.00	47.81	M	M	41.14	26.92	M	M	M	32.21	27.98	M	29.20	M
P6	66.80	57.01	53.15	M	M	48.02	47.67	M	M	38.23	37.88	M	M	34.02	34.37	28.89	M	M	19.10	M	15.24
	379.69	318.84	114.51	99.62	105.03	164.90	130.82	78.03	83.43	125.65	95.95	41.14	26.92	34.02	34.37	74.48	65.36	43.77	39.62	29.20	15.24

Figure 10.12   The product-core burden matrix.

No. of Cores	Core No.	Core Cost	Feasible Products	Product Cost	Total Cost
1	2	318.84	P1	52.86	318.84
			P2	40.18	
			P3	47.42	
			P4	56.14	
			P5	65.22	
			P6	57.01	
2	10	138.61	P2	34.37	182.38
			P3	28.64	
			P4	37.36	
			P6	38.23	
	18	43.77	P1	15.79	
			P5	27.97	
3	10	138.61	P2	34.37	197.62
			P3	28.64	
			P4	37.36	
			P6	38.23	
	18	43.77	P1	15.79	
			P5	27.97	
	21	15.24	P6	15.24	
4	10	138.61	P2	34.37	224.54
			P3	28.64	
			P4	37.36	
			P6	38.23	
	13	26.92	P5	26.92	
	18	43.77	P1	15.79	
			P5	27.97	
	21	15.24	P6	15.24	

Figure 10.13 Comparison when changing the number of cores.

accurate demand forecast environment and inaccurate demand forecast. The performance measures used to compare the different policies are the holding cost, manufacturing time, and lost sales cost. The comparison depends on the following assumptions:

1. General assumptions:
   The manufacturing capacity and resources are available in infinite scale
   Holding cost is 20% of product cost
   Every unsold item costs $2.5
   Manufacturing time represents the total time needed to manufacture the products on multiple production lines and multiple shifts

2. Make-to-stock assumptions:
   The demand is fulfilled at the end of the second and fourth week of every month
   The inventory should have all products forecasted available at the beginning of each month, and no manufacturing during a month to cover the demand of the same month
3. Make-to-order assumptions:
   Orders arrive at the beginning of the second and fourth week in every month
   No items are stocked in the inventory
4. Mass customization assumptions:
   The manufacturing operations start a week before the delivery date
   Number of cores to be stocked will be equal to the forecasted demand, and are available at the beginning of each month, and no manufacturing during a month to cover the demand of the same month.

10.3.3.1 Make-to-Stock vs. Mass Customization

Assume that there is no change in the following demand quantities shown in Table 10.4 for all products through a month.

According to the demand periods that are shown in Table 10.4, the demand holding periods for each product are shown in Figure 10.14. For the make-to-stock policy, the cost of a product is calculated based on manufacturing each product from a solid workpiece proposed in the previous section. The holding cost calculations are shown in Table 10.5 based on the equation, holding cost = average inventory × item cost × holding cost. There is no manufacturing time because the lead time is zero.

For the mass customization policy core A is Y_{10} and core B is Y_{18}. There is a ratio that should be consumed from each core by each product. Table 10.6

Table 10.4 Demand Quantities and Periods for Case Study Products

Product	Forecasted Demand	Demand Period
P1	100	Week 2
P2	150	Week 4
P3	80	Week 2
P4	150	Week 4
P5	350	Week 4
P6	300	Week 2

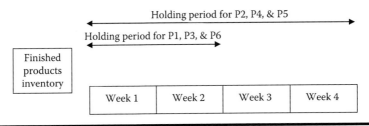

Figure 10.14   Demand holding periods for make-to-stock policy for each product shown in Table 10.4.

Table 10.5   Holding Cost Calculation for Make-to-Stock Policy

					Holding Cost Calculations
P	*Demand*	*Cost*	*Manufacturing Time (h)*	*Holding Cost ($)*	*Average Inventory × Item Cost × Holding Cost*
P1	100	70.2	0	702	$(100/2) \times 70.21 \times 0.2$
P2	150	57.5	0	1,726	$150 \times 57.54 \times 0.2$
P3	80	51.8	0	414	$(80/2) \times 51.8 \times 0.2$
P4	150	63.8	0	1,915	$150 \times 63.86 \times 0.2$
P5	350	69.4	0	4,862	$350 \times 69.46 \times 0.2$
P6	300	66.8	0	2,004	$(300/2) \times 66.8 \times 0.2$
Total			0	11,623	

Table 10.6   Product-Core Ratio

Product	*Forecasted Demand*	*Core*	*Product-Core Ratio (%)*
P1	100	B	22.2
P2	150	A	22.1
P3	80	A	11.7
P4	150	A	22.1
P5	350	B	77.8
P6	300	A	44.1

shows this product–core ratio. This ratio is defined as the product–core ratio. The demand of product P1 is 100 items, which is manufactured from core B. Core B is also used to manufacture P5, which has a demand of 350, so the product–core ratio for P1 is $(100/(100 + 350)) = (100/450) = 22.2\%$ and for P5 $(350/(100 + 350)) = (100/450) = 77.8\%$. This ratio is important to calculate the holding cost for each core. Figure 10.15 shows the holding periods for each core. The holding cost is calculated using the same formula for make-to-stock policy. Table 10.7 shows the manufacturing time and holding cost for each product for mass customization policy.

The manufacturing time is calculated based on the manufacturing time rates proposed in the previous section. Determining the number of production

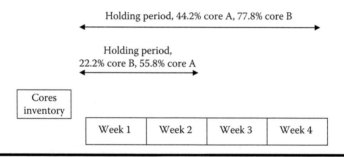

Figure 10.15　The holding periods for each core.

Table 10.7　Manufacturing Time and Holding Cost Calculation for Mass Customization Policy

Core	Period (Week)	Demand	Cost	Manufacturing Time (h)	Holding Cost ($)	Holding Cost Calculations — Average Inventory × Item Cost × Holding Cost
A	2	380	36.1	917	1373	$(380/2) \times 36.133 \times 0.2$
	4	300	36.1	717	2167	$300 \times 36.133 \times 0.2$
B	2	100	54.4	105	544	$(100/2) \times 54.42 \times 0.2$
	4	350	54.4	652	3809	$350 \times 54.42 \times 0.2$
Total				2391	7893	

Table 10.8   Forecasted Demand Deviation

Product	Forecasted Demand	Actual Demand	Demand Period
P1	100	80	Week 2
P2	150	120	Week 4
P3	80	64	Week 2
P4	150	165	Week 4
P5	350	385	Week 4
P6	300	330	Week 2

lines and shifts needed to manufacture each product is beyond the scope of this study.

Assume deviations in the demand from the forecasted quantities as shown in Table 10.8. For the make-to-stock policy, the holding cost calculation is shown in Table 10.9. For the mass customization policy, the manufacturing time and holding cost calculation is shown in Table 10.10.

10.3.3.2 Make-to-Order vs. Mass Customization

The same demand quantities that are shown in Table 10.4 are considered for make-to-order policy.

For the make-to-order policy, the orders receipt and due dates are shown in Figure 10.16. The required manufacturing times are shown in Table 10.11.

For the mass customization policy, the manufacturing time and holding cost calculations are shown in Table 10.12.

It is assumed that the same previous demand deviations that are shown in Table 10.8 occurred with make to order policy.

For the make-to-order policy, the manufacturing time is shown in Table 10.13. For the mass customization policy, the manufacturing time and holding cost calculations are shown in Table 10.14.

Table 10.15 shows a comparison of the results for make-to-stock, make-to-order, and mass customization policies under accurate demand. The results show that make-to-stock has the highest holding cost while make-to-order has the longest time to manufacture products with a holding cost equaling zero. Mass customization policy comes in as intermediate. Mass customization costs involves holding items in stores (less than make-to-stock policy) and requires time for manufacturing products (less than that needed in make-to-order policy). Figure 10.17 shows the position of mass customization with respect to

Table 10.9 Holding Cost Calculation for Make-to-Stock Policy with Deviated Demand

Product	Demand	Cost	Manufacturing Time (h)	Holding Cost ($)	Lost Sales ($)	Holding Cost Calculations — Average Inventory × Item Cost × Holding Cost	Remarks
P1	80	70.2	0	842	0	$[(80/2) \times 70.21 \times 0.2]$ + $[20 \times 70.21 \times 0.2]$	Holding 80 items for 2 weeks and 20 items for 4 weeks
P2	120	57.5	0	1,017	0	$[(120/2) \times 57.54 \times 0.2]$ + $[30 \times 57.54 \times 0.2]$	Holding 120 items for 2 weeks and 30 items for 4 weeks
P3	64	51.8	0	497	0	$[(64/2) \times 51.8 \times 0.2]$ + $[16 \times 51.8 \times 0.2]$	Holding 64 items for 2 weeks and 16 items for 4 weeks
P4	165	63.8	0	1,915	37.5	$150 \times 63.86 \times 0.2$	15 items lost sales
P5	385	69.4	0	4,862	87.5	$350 \times 69.46 \times 0.2$	35 items lost sales
P6	330	66.8	0	2,004	75	$(300/2) \times 66.8 \times 0.2$	30 items lost sales
Total			0	11,137	200		

Table 10.10 Manufacturing Time and Holding Cost Calculation for Mass Customization Policy with Deviated Demand

Core	Period (Week)	Demand	Cost	Manufacturing Time (h)	Holding Cost ($)	Lost Sales ($)	Holding Cost Calculations: Average Inventory × Item Cost × Holding Cost	Remarks
A	2	394	36.1	917	1373	35	$(380/2) \times 36.133 \times 0.2$	14 items lost sales
	4	285	36.1	648	2167	0	$300 \times 36.133 \times 0.2$	Holding extra 15 items for 4 weeks
B	2	80	54.4	84	544	0	$(100/2) \times 54.42 \times 0.2$	Holding extra 20 items for 2 weeks
	4	385	54.4	652	3809	87.5	$350 \times 54.42 \times 0.2$	35 items lost sales
Total			2301	7893	122.5			

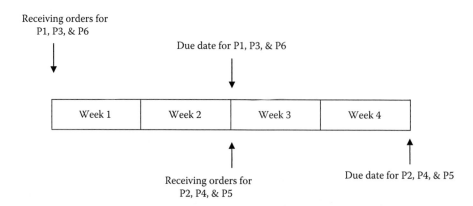

Figure 10.16 Orders receipt and due dates for make to order policy.

Table 10.11   Required Manufacturing Times for Each
Product for Make-to-Order Policy

Product	Demand	Cost	Manufacturing Time (h)	Holding Cost ($)
P1	100	70.2	468	0
P2	150	57.5	575	0
P3	80	51.8	276	0
P4	150	63.8	638	0
P5	350	69.4	1620	0
P6	300	66.8	1335	0
Total			4912	0

make-to-stock and make-to-order policies regarding the holding cost and the manufacturing time.

The holding cost for mass customization policy is lower than that for make-to-stock by 32%. The time that is needed in mass customization for manufacturing products is lower by 51% than that for make-to-order. This means that the time that a customer is willing to wait in mass customization is half that for make-to-order policy. Table 10.16 shows a comparison between make-to-stock and mass customization manufacturing policies under inaccurate demand. The comparison shows that when the demand is inaccurate, the holding cost for mass customization is lower than that for make-to-stock by 29% and the lost sales cost is lower by 38%. But in mass customization, a wait time of 2303 h is spent as manufacturing

Table 10.12   Manufacturing Time and Holding Cost Calculations for Mass Customization Policy

Core	Period (Week)	Demand	Cost	Manufacturing Time (h)	Holding Cost ($)	Holding Cost Calculations
						Average Inventory × Item Cost × Holding Cost
A	2	380	36.1	917	1373	$(380/2) \times 36.133 \times 0.2$
	4	300	36.1	717	2167	$300 \times 36.133 \times 0.2$
B	2	100	54.4	105	544	$(100/2) \times 54.42 \times 0.2$
	4	350	54.4	652	3809	$350 \times 54.42 \times 0.2$
Total				2391	7893	

Table 10.13   Manufacturing Time for Make-to-Order Policy with Deviated Demand

Product	Demand	Cost	Manufacturing Time (h)
P1	80	70.2	84
P2	120	57.5	275
P3	64	51.8	221
P4	165	63.8	702
P5	385	69.4	1782
P6	330	66.8	1469
Total			4533

time but in make-to-stock, the customer can buy the product without waiting for manufacturing. If a company is interested in lower cost, the suitable policy is mass customization. On the other hand, if a company is interested in delivery date, make-to-stock policy is more suitable. The comparison is represented in a three-dimensional graph shown in Figure 10.18.

Table 10.17 shows a comparison between make-to-order and mass customization manufacturing policies under inaccurate demand. Table 10.17 shows that in an inaccurate demand environment, there is no lost sales cost for either make-to-order or mass customization policy because manufacturing is accomplished when

Table 10.14 Manufacturing Time and Holding Cost Calculation for Mass Customization Policy with Deviated Demand

Core	Period (Week)	Demand	Cost	Manufacturing Time (h)	Holding Cost ($)	Holding Cost Calculations — Average Inventory × Item Cost × Holding Cost	Remarks
A	2	394	36.1	917	1373	$(380/2) \times 36.133 \times 0.2$	14 items lost sales
	4	285	36.1	686	2167	$300 \times 36.133 \times 0.2$	Holding extra 15 items for 4 weeks
B	2	80	54.4	84	544	$(100/2) \times 54.42 \times 0.2$	Holding extra 20 items for 2 weeks
	4	385	54.4	718	3809	$350 \times 54.42 \times 0.2$	35 items lost sales
Total				2405	7893		

Table 10.15   Comparison among Manufacturing Policies under Accurate Demand

Accurate Demand			
	Make-to-Stock	*Make-to-Order*	*Mass Customization*
Total holding cost ($)	11,623	0	7,894
Total manufacturing time (h)	0	4,914	2,392

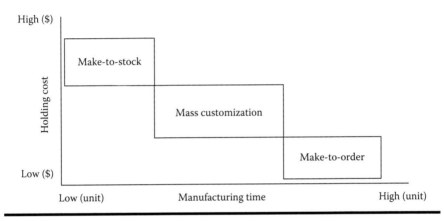

Figure 10.17   Mass customization with respect to make-to-stock and make-to-order policies regarding the holding cost and the manufacturing time.

Table 10.16   Comparison between Make-to-Stock and Mass Customization Manufacturing Policies under Inaccurate Demand

Inaccurate Demand Forecast		
	Make-to-Stock	*Mass Customization*
Holding cost ($)	11,137	7,893
Lost sales ($)	200	122.5
Manufacturing time (h)	0	2,301

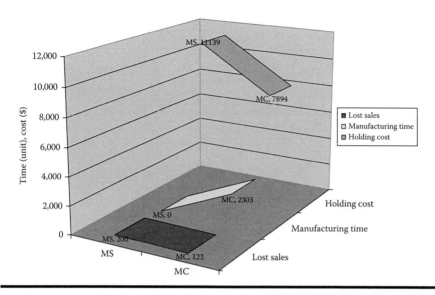

Figure 10.18 Make-to-stock vs. mass customization under inaccurate demand forecast conditions.

Table 10.17   Comparison between Make-to-Order and Mass Customization Manufacturing Policies under Inaccurate Demand

Inaccurate Demand Forecast		
	Make-to-Order	*Mass Customization*
Holding cost ($)	0	7893
Lost sales ($)	0	0
Manufacturing time (h)	4533	2405

receiving an order and upon the request of a customer, and the company can cover all possible demand. For mass customization policy, there is a holding cost, and a customer should wait for the demand delivery. The time needed for waiting is less than that for make-to-order policy. Figure 10.19 shows a three-dimensional representation of the comparison between make-to-order and mass customization under inaccurate demand forecast conditions.

The comparison shows that mass customization policy comes as intermediate between make-to-stock and make-to-order. This implies that mass customization has the advantages of low holding cost and low manufacturing time. For mass customization, the effect of inaccurate forecasted demand is the lowest compared to make-to-stock and make-to-order policies. On the other hand, a customer should wait for manufacturing time to obtain the product.

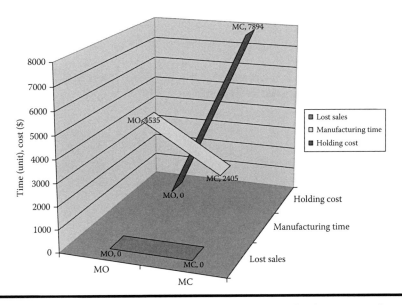

Figure 10.19 Make-to-order vs. mass customization under inaccurate demand forecast conditions.

10.4 Conclusions

A model to efficiently implement mass customization at the manufacturing level was developed. The model consists of two phases. The first phase begins by analyzing product's manufacturing features and ends with identifying starting features. The second phase is called manufacturing cores formation. This phase identifies the optimum core(s) to use from a set of possible cores that are identified. This approach provides certain advantages such as stocking cores (semifinished products), which are used to manufacture a range of products after performing certain operations. This means that the approach can cope with the differences between demand and forecast. Also, stocking semifinished products rather than finished products lowers the holding cost. The presented approach comes as intermediate production strategy between make-to-stock and make-to-order strategies. It depends on the managerial decision to identify which strategy will be addressed because each strategy has its own advantages.

Using manufacturing cores enables organizations to differentiate products from lower number of subassemblies because a manufacturing core is considered to be a modular product from a design point of view. Manufacturing cores allow stocking a lower number of subassemblies and provide more flexibility under inaccurate demand forecast. But, on the other hand, the model complexity will increase as the number of products and generic and variant features increase, making it very difficult to implement; thus the model could be supplemented with an automated feature recognition module or subsystem to facilitate the extraction of manufacturing features. Besides the optimum number of cores, N_p was determined by solving the optimization model several times, each with a different preset number. This made the optimization somewhat iterative and cumbersome. Thus, the optimization should be extended or modified to optimize for the number of cores directly.

Authors

Hazem Smadi has enrolled in the PhD program at the Industrial Engineering Department at the University of Houston. His research area is manufacturing systems, CAD/CAM, product design, and rapid prototyping.

Ali K. Kamrani is an associate professor of industrial engineering, and director of the industrial engineering graduate program studies and accelerated BS to MS programs. He is also the founding director of the Design and Free Form Fabrication Laboratory at the University of Houston, United States. He has been a visiting professor at the Princess Fatimah Alnijris's Research Chair for AMT, Industrial Engineering Department at King Saud University, Riyadh. Saudi Arabia. He received his BS in electrical engineering in 1984, his MEng in electrical engineering in 1985, his MEng in computer science and engineering mathematics in 1987, and his PhD in industrial engineering in 1991, all from the University of Louisville, Louisville, Kentucky. His research interests include the fundamental application of systems engineering and its application in the design and development of complex systems. He is the editor in chief for the *International Journal of Collaborative Enterprise* and the *International Journal of Rapid Manufacturing*. He is a professional engineer at the State of Texas.

Sa'Ed M. Salhieh is an associate professor of industrial engineering at the University of Jordan. He specializes in new product development. He is currently the director of the Product Design Laboratory at the University of Jordan.

Dr. Salhieh has a BSc in mechanical engineering from Jordan University of Science and Technology; an MSc in industrial and systems engineering from the University of Michigan, Dearborn; and a PhD in industrial engineering from Wayne State University. His primary research interests are modular design, collaborative engineering, product development, and information modeling. Dr. Salhieh's research activities include developing novel solution methodologies to enhance the product development process.

His research interests includes investigating new avenues to address the fuzzy front end of the product design process. Currently, he is planning to develop methodologies using real options analysis and logical gates to address the problem of selecting new product concepts.

References

Berman, B. 2002. Should your firm adopt a mass customization strategy? *Business Horizons*. 45: 51–60.

Gu, X., Qi, G., Yang, Z., and G. Zheng. 2002. Research of the optimization methods for mass customization (MC). *Journal of Materials Processing Technology*. 129: 507–512.

Jiao, J., Ma, Q., and M. Tseng. 2003. Towards high value-added products and services: Mass customization and beyond. *Technovation*. 23: 809–821.

Li, W., Ong, S., and A. Nee. 2002. Recognizing manufacturing features from a design-by-features model. *Computer Aided Design*. 34: 849–868.

Radder, L. and L. Louw. 1999. Mass customization and mass production. *The TQM Magazine*. 11: 1–35.

Silveira, G., Borenstein, D., and F. Fogliatto. 2001. Mass customization: Literature review and research directions. *International Journal of Production Economics*. 72: 1–13.

Smadi, H. 2005. Product Customization through the Development of Manufacturing Cores. Masters Thesis. Industrial Engineering Department, University of Jordan, Amman, Jordan.

Tseng, M. and J. Jiao. 2001. *Mass Customization, Handbook of Industrial Engineering*. New York: John Wiley & Sons.

Tseng, M. and J. Jiao, 2004. Customizability analysis in design for mass customization. *Computer Aided Design*. 36: 745–757.

Yang, S. and T. Li. 2002. Agility evaluation of mass customization product manufacturing. *Journal of Materials Processing Technology*. 129: 640–644.

Chapter 11

Genetic-Algorithm-Based Solution for Combinatorial Optimization Problems

Ali K. Kamrani

Contents

This chapter presents various methods and algorithms used in the solution of combinatorial optimization problems. First, a definition of a combinatorial optimization problem is given, followed by a discussion on the depth-first branch-and-bound algorithm and the local search algorithm, two approaches used for solving these kinds of problems. An introduction to genetic algorithms (GAs) coupled with an explanation of their role in solving combinatorial optimization problems is then presented. The discussion on GAs focuses on the advantages and disadvantages of using this technique as a tool for solving combinatorial optimization problems. Finally, some conclusions are presented to emphasize the implications, benefits, and drawbacks of using GAs in the solution of various combinatorial optimization problems. A sample case in modular design is also presented.

11.1 Introduction

The family of combinatorial optimization problems is characterized by having a finite number of feasible solutions. These problems abound in everyday life, particularly in engineering design. In principle, finding the optimal solution for a finite problem could be done by simple enumeration. However, real-life problems are much more complicated and enumeration is frequently an impossible technique to use because the number of feasible solutions can be enormous. Combinatorial optimization problems are derived from combinatorics, a branch of mathematics concerned with the problem of arranging and selecting discrete objects. However, as Law (1976) points out, while combinatorics tries to find out if a particular arrangement of objects exists in a solution set, combinatorial optimization problems go a step further, trying to determine if a given arrangement can be an optimal solution to the problem at hand.

Several methods and algorithms have been used in the solution of various combinatorial optimization problems. Sait and Youssef (1999) divided them into two major categories: exact algorithms and approximation algorithms. The first category includes linear programming methods, dynamic programming, and branch and bound, among many others. Linear programming approaches formulate the problem at hand as either maximization or minimization of a certain objective subject to a number of constraints. Dynamic programming is a search method suitable for optimization problems whose solution can be obtained as the result of a sequence of decisions or steps. Branch and bound searches the solution space tree, trying to find the optimal solution to the problem.

Most exact algorithms have the common problem that in nature, they all are enumerative, a fact that creates a problem when one tries to solve a real-life problem that has a lot of constraints and difficulties. As was mentioned earlier, these problems can have a very large number of solutions, and simply enumerating them and finding the best one is not efficient and sometimes impossible. Approximation algorithms are a partial solution to the problems found with enumeration techniques. They constitute the second category of algorithms discussed by Sait and Youssef (1999). Usually known as heuristic methods, these algorithms give a viable option when trying to solve very complex problems. They search only a portion of the solution space heuristically and find good solutions to a problem based on a number of constraints. These techniques are more efficient, as they give solutions "faster" than the enumeration methods discussed before. However, as it will be seen, there is a trade-off in using heuristics instead of simple enumeration techniques. The trade-off consists in "giving up" the possibility of getting an optimal solution in order to achieve acceptable results in a reasonable amount of time. Some examples of approximation algorithms include the constructive greedy method, local search, simulated annealing, genetic algorithms (GAs), tabu search, stochastic evolution, and simulated evolution. This chapter focuses on GAs as a way of solving difficult combinatorial optimization problems.

GAs represent a great alternative for solving a wide range of difficult optimization problems. They have shown great power with very promising results in experimentation and practice of many industrial engineering areas. GAs use meta-heuristics to solve these problems. These heuristics provide the user great flexibility because he orshe does not have to have great problem-specific knowledge in order to arrive at good solutions. In addition, these heuristics are very general, as a GA can be used to solve almost any combinatorial optimization problem without making many modifications to the algorithm or to the problem itself. However, as was mentioned earlier, these heuristics could generate some problems due to the fact that GAs are blind in nature (i.e., they do not know when they have reached an optimal solution). Also, GAs can give the user a solution, but there is no guarantee that it will be the optimal one. By definition, the heuristics used by GAs and simulated annealing, among other algorithms, are approximations, so one can never be sure if the solution found by these algorithms is an optimal one. Some classic examples of combinatorial problems include the knapsack problem, the quadratic assignment problem, the minimum spanning tree problem, the traveling salesman problem (TSP), and the film-copy deliverer problem. All these problems have been solved (or partially solved) using GAs.

The rapid growth of the computer industry, along with the expanding power that computers possess, has increased the research interest in many engineering fields that try to solve very complex combinatorial optimization problems. Tian et al. (1999) investigated the effects of using simulated annealing on several problems such as the TSP, the flow-shop scheduling problem, and the quadratic assignment problem. Ahuja et al. (1999) suggested and tested a "greedy" GA for the quadratic assignment problem. Breedam (1999) used tabu search and simulated annealing to solve a vehicle routing problem. Chu et al. (2000) used GAs to improve and create new network designs that are more reliable and secure in this technology-oriented era. Ishibuchi et al. (1995) proposed a GA method for a classification system using if-then-else rules.

This chapter reviews various methods and algorithms used in the solution of combinatorial optimization problems. A discussion is made specifically on two methods that fall into the graph theory category: the enumeration technique known as branch and bound and the heuristic (approximate) method known as local search. An introduction to GAs coupled with an explanation of their role in solving combinatorial problems such as the TSP is also provided. A proposed solution to the TSP using a GA is presented. Finally, some conclusions are presented to emphasize the benefits and drawbacks of using GAs in various optimization problems.

11.2 Combinatorial Optimization Problems

Combinatorial optimization problems are getting harder every day. Today's designers have to develop and optimize complicated designs that must fit the manufacturing system and this should be doable in a reasonable amount of time. Airlines are introducing more and more flights to service their customers better, and companies are trying to reduce handling times and transportation times to decrease their costs and gain more profits. As the industry grows, so does the complexity of the problems it experiences. Many of these problems can be classified as combinatorial optimization problems. How do we minimize the material handling costs? How do the schedule our flights to serve our customers better? These are just a few questions managers are asking themselves on a daily basis. Their questions can be answered in many ways, but a reliable method that has been used to solve and understand these difficult problems is the study of graph theory. A generic model of an linear integer combinatorial problem is shown below (Hoffman, 2000):

$$\max \sum_{j \in N} c_j x_j + \sum_{j \in I} c_j x_j + \sum_{j \in C} c_j x_j \tag{11.1}$$

subject to

$$\sum_{j \in N} a_{ij} x_j + \sum_{j \in I} a_{ij} x_j + \sum_{j \in C} a_{ij} x_j \le \text{ or } \ge \text{ or } = b_i \tag{11.2}$$

$$l_j \le x_j \le u_j \quad (j \in I \cup C)$$

$$x_j \in \{0,1\} \quad \text{for} \left(j \in N \right) \tag{11.3}$$

$$x_j \in Z \quad \text{for} \left(j \in I \right) \tag{11.4}$$

$$x_j \in Y \quad \text{for} \left(j \in C \right) \tag{11.5}$$

N, I, and C are sets of 0–1, integer, and continuous variables. The l_j and u_j are the lower and upper bound values for the variable x_j. The problem is a pure 0–1 linear programming problem if C and I are equal to an empty set. If C is the empty set, it is a pure integer (linear) programming problem, otherwise it is a mixed integer (linear) programming problem. Zhang (1999) summarized a number of methods and algorithms used in the solution of various combinatorial problems. They all use tree structures in their search techniques, one of the basic elements of graph theory. Some of these methods include best-first search, iterative deepening, recursive best-first search, space-bound best-first search, depth-first branch and bound, and local search. This chapter discusses two methods derived from graph theory to solve combinatorial optimization problems.

11.2.1 Exact Method: Depth-First Branch and Bound

Depth-first search is one of many search techniques used by the branch and bound method. The underlying idea of this method is to take a given problem that could be difficult to solve directly, and decompose it into smaller sub problems is a way that a solution to the sub problem is also a solution to the overall problem. Depth-first searches for solutions in the tree by keeping track of the cost of the best solution found so far as an upper bound u that restricts the search space and reduces the time it takes to find a solution to a problem. Depth-first branch and bound selects the most recently generated node or the deepest node to expand next. If a new leaf node is reached, and its cost is less than u, the value of u is modified to reflect this new minimum cost. If, on the other hand, the cost of the node is greater than u, the node is pruned, or disregarded from further consideration. The step-by-step algorithm as described by Zhang (1999) is showed in the following:

Algorithm Depth-firstBranch&Bound (n)
Generate all k children of n
Evaluate cost of each child
For i = 1 to k
   If (cost(n$_i$) < u) *then*
      If (n$_i$ is a leaf node) *then*
         U = cost(n$_i$)
      Else Depth-firstBranch&Bound (n$_i$)
   Next I
End

A few researchers interested in solving linear combinatorial optimization problems have used the depth-first branch and bound method. Mans et al. (1995) proposed a depth-first branch and bound algorithm for solving the quadratic assignment problem. Sridhar and Chandrasekharan (1995) compared various search techniques used in many combinatorial problems. Pemberton and Weixiong (1996) implemented and tested a modification of the algorithm shown above to solve the TSP. They found this algorithm to be a viable option when trying to solve hard combinatorial optimization problems.

This method is, as the title of the section implies, an exact method for getting a solution of a linear combinatorial optimization problem. At the end of the search, there is a path that represents the choices made from a starting point (the root node) to an end point (a leaf node). In addition, the end value of u represents the cost of the path, which is the minimum possible cost incurred in the solution of this problem. As problems get bigger and more complex, regular search techniques and methods for finding optimal solutions become almost useless. Their complexities are high, so as problems grow, the computational costs grow as well, and the time it takes to find a solution increases rapidly. To overcome these problems, new techniques have been developed and tested. They all try to reduce complexity and in fact, are able to find solutions in shorter periods of time. However, as it will be mentioned, these methods do not always find optimal solutions, they just find good solutions in reasonable amounts of time.

11.2.2 Approximation Method: Local Search

Local search is a highly adaptive approximation method that has been used in a variety of combinatorial optimization problems. In fact, most optimization problems can be solved using this technique. Despite the fact that local search cannot specify if the best solution found so far is optimal, this method has had great success in solving large and complex combinatorial optimization problems. Local search starts with an initial feasible solution that will be improved by searching what Sait and Youssef (1999) and Zhang (1999) called the "neighborhood" of this initial solution. If in fact a better solution is found, and the optimal solution is updated by this new discovery, the algorithm will start the search again by searching the neighborhood of the latest solution until it finds the local optimum. The following pseudocode shows an outline of the step-by-step algorithm used by Sait and Youssef (1999). Local search can be thought of as a random walk in the solution space. It starts at a root node, which gets expanded in an attempt to reach for new nodes that are close, or in the neighborhood of this initial node or solution. Local search then repeatedly moves the current node to one of the neighboring nodes to find better solutions than the best one encountered so far. There are a lot of strategies on how these moves should be performed, and they play a major role in the performance of the algorithm. Inadequate strategies will cause the algorithm to get "lost" in local optimal solutions rather than in overall optimal solutions.

Algorithm LocalSeach (S_0)
$S_1 = S_0$
Repeat
  $S_2 = S_1$
  $S_1 = $ **Improve** (S_2)
Until $S_1 = $ null
Return S_2
End

The **Improve** procedure returns null when there is no improvement between solutions (Local optimum is found).

These strategies are represented in above pseudocode as the *Improve* subroutine. As it implies, this procedure will stop whenever it finds an optimal solution, even if the solution is not the global optimum. Even with very clever heuristics, it is very difficult for the algorithm to be sure that it has found the best possible solution for a problem. This is a trade-off the user usually has to accept, as local search gives solutions much faster than regular search techniques do. Moreover, the algorithm's complexity is smaller, giving the user greater flexibility on the size constraints of the problem. Local search has been widely used by many researchers in a wide range of fields or applications. Voudouris and Tsang (1999) examined the effects of using local search in the solution of the TSP. Angel and Zissimopoulos (2000) tested the efficiency of local search, simulated annealing, and tabu search in the solution of NP-hard optimization problems. Rios and Bard (1998) implemented two heuristic algorithms using local search to solve the flowshop scheduling problem. Local search algorithms provide users with good and fast solutions to hard combinatorial optimization problems. Although they are blind in nature, and there is no way to specify if a certain solution is just a local optimum, local search is one of the few efficient options available when trying to solve complex optimization problems. Along with local search, simulated annealing, tabu search, and GAs provide the best alternatives for solving these difficult problems. All these algorithms are blind as well, and so, there is no real way to know if a solution is the optimal one. However, as is explained in the next section, GAs try to reduce the probability of getting optimal solutions that are just local by expanding the search space. Nevertheless, GAs experience the same problems, and thus the initial solution given to the algorithm should be a clever one. Otherwise the probability of getting a good solution will be the same while the probability of getting *the optimal solution* will decrease dramatically.

11.3 Genetic Algorithms

Many combinatorial optimization problems from the manufacturing systems world are very complex and hard to solve using conventional optimization techniques. As it has been suggested throughout this chapter, enumeration techniques are usually obsolete if a problem is extremely big. Since real-life problems are in most cases enormous, new techniques have to be developed and used to solve these large and difficult problems. Many techniques have been proposed since the development of the local search algorithm discussed in the previous section. Some techniques include simulated annealing, tabu search, and GAs.

The rapid development of computer science, along with the rapid development of artificial intelligence techniques, has created a great interest among engineers in imitating living beings to solve those kinds of difficult problems. Michalewicz (1992) explained that the simulation of the natural evolutionary process of human beings results in stochastic optimization techniques called evolutionary algorithms. The idea is that by simulating the thought process that the human uses to solve difficult problems, the algorithm will be able to solve the same kind of problems. Moreover, and unlike the normal local search algorithm, evolutionary algorithms (and more specifically GAs) will try to take into account a wider range of possible solutions, which will then in turn reduce the probability of not finding an optimal solution to a given combinatorial problem.

Currently, there are three areas of research in the field of evolutionary algorithms: GAs, evolutionary programming (EP), and evolution strategies (ESs). GAs are by far the most widely known and used method for solving combinatorial optimization problems. Some applications of GAs in the manufacturing world include scheduling and sequencing, group technology, reliability design, vehicle routing and scheduling, facility layout and location, transportation, among many others. Note that all these applications can generate a wide variety of hard combinatorial optimization problems that can only be solved with highly adaptive algorithms in a reasonable amount of time. Many companies have used GAs to solve various problems that are present within these applications. Researchers have focused their attention on several difficult combinatorial optimization problems such as the knapsack problem, the quadratic assignment problem, and the TSP. Grefenstette et al. (1987) solved the TSP using GAs. Min and Cheng (1999) proposed a GA for minimizing the makespan in identical machine scheduling. Tate and Smith (1995) presented a GA that solved the quadratic assignment problem. They stressed the importance of giving the GA a good initial solution as a way of improving the performance of the algorithm. Dellaert et al. (2000) developed a GA approach to solve a general multilevel lot-sizing problem. Their results show a significant improvement in time and cost, which enables users to solve much more complex problems without disregarding important facts about the problem.

Holland (1975), one of the pioneers in the area of evolutionary algorithms, stated that GAs are stochastic search techniques based on the mechanism of natural selection and natural genetics. They start with a set of random solutions called population. Each individual or solution in the population is called chromosome. As Goldberg (1989) explained, a chromosome is a string of symbols (usually, a binary bit string), which represents a solution to the problem being solved and discussed. The initial population has to be determined by the user. Each combinatorial

optimization problem is different, so special attention has to be given to the definition of a solution. Solutions can be represented as binary bit strings, number
strings, word strings, among many other options. However, if a representation does
not accurately represent what the solution actually means, the algorithm will perform poorly, and no useful information will be obtained. In essence, the algorithm
will search for an optimal solution in an inaccurate solution space. Chromosomes
evolve via successive iterations called generations. During each iteration, the chromosomes are evaluated using a fitness function that will eventually decide if a chromosome passes to the next generation or dies. After an iteration of the algorithm, a
new generation is created with new chromosomes called offspring. They are formed
by either merging two chromosomes from the current generation using the crossover operation or by modifying a single chromosome using the mutation operation.
The new generation is then formed by selecting some of the chromosomes according to their fitness value and by rejecting others that do not qualify as valuable
individuals.

Fitter chromosomes are more likely to live longer. In essence, they have a higher
probability of survival. According to Holland (1975), after a few runs (a few generations), the algorithm should converge to the best chromosome, which hopefully
represents the optimal solution to the problem. However, as many researchers have
stressed, this solution may not be the optimal one. This is due to the fact that the
algorithm may be "lost" in an area that does not have the best feasible solution
and/or the initial solution given by the user does not accurately represent a possible
solution to the problem at hand. Figure 11.1 illustrates the basic process of a GA.

In this case, the original population is composed of four individuals, which
are divided into two groups, forming two pairs of "parents." Parents experience
crossover (in this case, single-point crossover, which will be explained later) and
become new chromosomes (called offsprings). After crossover, one of the chromosomes undergoes mutation (one of its attributes changes from 0 to 1). When the
crossover and mutation operations have done their job, a fitness function is applied
to the chromosomes to determine the fitness of each individual. In this case, the
fitness value is determined by the number of ones on a chromosome. The usage of
these values determines which chromosomes will survive and which chromosomes
will die. In this case, the inferior individual (the one that has a fitness of 2) is
eliminated from the population and the superior individual (the one with fitness
6) is duplicated. This new population is then tested. If the population is not good
enough, another set will be created and the process is then repeated until a good
population is found (in essence, the creation of new generations mimics the evolutionary process). At this point, an optimal solution is found and no more iterations
are needed, so the process is terminated.

Obitko (1998) summarized a number of advantages and disadvantages that
GAs can bring to any kind of optimization problem, stressing their importance as
an efficient tool for solving hard optimization problems. GAs are relatively easy to
develop and validate, because as mentioned earlier, they are very adaptive and the

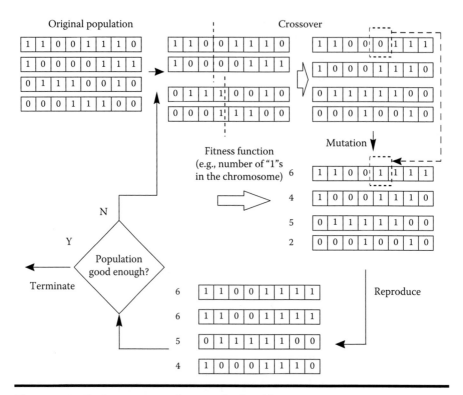

Figure 11.1 Basic structure of a genetic algorithm.

user does not have to make many changes to an already useful and proved algorithm in order to get reasonable solutions to a new optimization model. Michalewicz (1992) and Wang (1998) also mentioned the efficiency of this method by showing the advantages of working with a population of solutions rather than with a single solution. GAs are capable of introducing many new solutions after each iteration, an advantage that local search, for example, does not have. The neighborhood of a solution is searched more efficiently and more thoroughly because more possibilities are explored.

Another advantage of applying GAs to combinatorial optimization problems is that they do not have a lot of mathematical requirements about the optimization problem. Given their evolutionary nature, GAs search for solutions without regard to the specific inner workings of the problem. GAs can handle any kind of objective functions and any kind of constraints. The last advantage of using GAs for combinatorial optimization problems can also be classified as their first disadvantage. As has been stressed many times, the use of GAs reduces the possibility of finding local optimal solutions rather than the global optimal one. However, this possibility does not disappear, and so the user has to be aware that the algorithm will not necessarily give the optimal solution every time. This disadvantage surfaces another problem that GAs have (in fact, all approximation algorithms have it). GAs are blind in nature, so there is no real way of knowing if a solution given by the algorithm is the best one or just a good solution. Also, the algorithm does not really know if it has arrived at an optimal solution or not (i.e., the algorithm cannot tell if a given solution is optimal). Moreover, and as Wall (1995) mentioned, the solutions have to be coded so that the algorithm can work with them, and then decoded so that the user understands what the algorithm is trying to say. This process can be long and its accuracy is critical. Finally, the cost of running a GA to solve a combinatorial optimization problem can be very high (Michalewicz, 1992). Many iterations and generations have to be searched, and so the space used and needed by the algorithm is usually greater than the space required by a regular optimization technique. Moreover, a larger number of iterations are performed (in order to take full advantage of the benefits of a GA), an event that can prove this method useless because of the time constraints imposed to get good and accurate solutions.

Although there can be a lot of disadvantages associated with GAs and their use as tools for solving combinatorial optimization problems, the benefits one can get from using them are substantial. Along with the advantages previously mentioned, GAs try to mimic the approach a human would take when trying to solve a problem. This is highly beneficial because the experience humans bring to a new situation can be simulated by allowing the algorithm to search a broader field of solutions rather than just following a single path in a tree. Holland described the basic structure of a GA in 1975. This structure follows the same basic process shown in Figure 11.1. The crossover operation used by Holland (1975) was the single-point approach, and the selection probability of chromosomes for reproduction is determined by the chromosome's fitness value. Based on Holland's algorithm, Obitko (1998) summarized the algorithm in his work. The algorithm shown below is highly adaptive, and many combinatorial optimization problems can and have been solved using this basic outline:

Procedure: *Genetic Algorithm*

1. **Initialization:** Generate random population of n chromosomes.
2. *Evaluation*: Evaluate the fitness $f(x)$ of each chromosome x in the population.
3. **New population:** Create a new population by repeating the following steps until the new population is complete:

   Selection: Select two parent chromosomes from a population according to their fitness (the better fitness, the bigger chance to be selected).

   Crossover: Given a crossover probability, crossover the parents to form new offspring. If no crossover is performed, offspring is an exact copy of parents.

   Mutation: With a mutation probability, mutate new offspring at given positions in a chromosome.

   Accepting: Place new offspring in a new population.

4. **Replace:** Use new generated population for a further run of algorithm (new generation).
5. **Test:** If the end condition is satisfied (usually, required fitness value is obtained), STOP, and return the best solution in current population.
6. **Loop:** Otherwise go to step 2 (a new generation will be created).

The initialization that creates the first set of solutions (first generation) can follow any technique, but it is usually a random process that creates strings of numbers, each of which represent a possible solution to the problem at hand. The initialization process, however, is not perfect. A random string of numbers may have all the basic requirements needed in a chromosome, but the combination of these numbers may result in the creation of an illegal solution. Many solutions have been developed to solve the initialization problem. Cheng and Gen (1997) discussed and divided these methods into two major categories: the penalty method and the decoder method. Gordon and Whitley (1992) gave a simple penalty for each infeasible solution that was created by the algorithm every time a new chromosome was created. Olsen (1993) examined three kinds of penalty methods for the knapsack problem. The idea of his penalty method was that every time an infeasible solution was found, a penalty was given to it, and this penalty would prevent the solution from ever being picked as a parent for a new generation. In essence, after the first generation, these infeasible solutions would die.

Gordon and Whitley (1992) also developed the decoder method, which generates a solution from a chromosome using the greedy approximation heuristic. This heuristic is essentially a blind strategy, because at each step of the way, the heuristic chooses the best alternative possible for that single step, which, in some cases, may lead to a nonoptimal solution. The greedy strategy has been used to solve the initialization problems for the knapsack problem and the TSP with some promising results, but this is a very problem dependent heuristic since most problems will not find an optimal solution using this kind of strategy. For combinatorial optimization problems, the evaluation process needs to be carefully defined. The performance of the fitness function is crucial for the algorithm's performance. If the evaluation technique is too general, the algorithm may never reach an optimal solution because the fitness function will not measure the chromosomes properly. On the other hand, if the process is too strict, the algorithm will be unable to find any kind of optimal solution. It will just run forever trying to satisfy some impossible conditions. This may be very costly if one takes into account the cost associated with the use of GAs in combinatorial optimization problems. The three most important steps in a GA will now be presented. These steps, coupled with the initialization and evaluation of the solutions, give GAs an edge in the solution of hard combinatorial optimization problems. They introduce new information that would not be otherwise searched and inspected by regular optimization methods, a benefit that can prove vital in the use of GAs as plausible optimization techniques.

11.3.1 Genetic and Evolution Operations

The initialization and evaluation processes are very important in the performance of a GA during the solution of a combinatorial optimization problem. However, the three steps that really measure the algorithm's performance are the crossover and mutation steps, and the selection step. These three steps can be divided into two kinds of operations that are performed within any GA:

1. Genetic operations: crossover and mutation
2. Evolution operation: selection

The genetic operations mimic the process of heredity of genes to create new offspring at each generation, while the evolution operation mimics the process of Darwinian evolution to create populations from generation to generation. Both operations introduce a wide range of possibilities to the problem being studied. By mixing and mutating existing chromosomes in the solution space, the algorithm explores solutions to the combinatorial problem that would never be explored by a regular optimization technique.

These operations are explained in the next subsections. The discussion focuses on the operations used by Cheng and Gen (1997) in their efforts for finding a solution to the TSP. A sample TSP is shown at the end of the section. The GA used for this problem is based on the approach discussed by Cheng and Gen (1997). Crossover may be the most important operation performed by a GA for the solution of a combinatorial optimization problem. It usually operates on two chromosomes at a time and generates offspring by combining them. A cut-point is randomly chosen and the segments are combined to create two new chromosomes. Cheng and Gen (1997) explained that the cut point can be one or many. If the crossover operation follows a one cut-point approach, then both chromosomes are mixed together: the "head" of one is paired with the "tail" of the other one and so two new chromosomes are created (check by looking back at Figure 11.1). If there is more than one cut point, the new chromosome will be an intercalated version of both parents.

This crossover approach was first denominated by Holland et al. (1986) as the canonical approach. In the TSP, this approach may lead to solutions that do not include some cities, or that may repeat several cities. This is highly undesirable, as these solutions cannot be accepted. Cheng and Gen (1997) embedded a repairing procedure in their algorithm to resolve this problem. In the sample TSP shown, replacing cities that may be repeated for cities that were left out of the solution repairs the illegality of the tours or solutions. However, the canonical approach is blind, a characteristic that may be very costly in terms of the repair procedure and the time it will take for the algorithm to find a solution to the combinatorial optimization problem (and more specifically to the TSP). Many researchers have developed and used various crossover techniques in the solution of a wide range of combinatorial optimization problems. Solutions to several of these problems may involve strange situations that must be resolved if the GA is to perform well. Cheng and Gen (1997) studied a simplification of the crossover operation that would work on chromosomes of unequal length. This simplification is called injection crossover and was developed by Falkenauer and Delchambre (1992) when they realized that in several combinatorial optimization problems such as the knapsack problem or the minimum spanning tree problem, solutions (or chromosomes) could vary in length due to the nature of the problem. Tate and Smith (1995) developed a crossover operation for the quadratic assignment problem. Their method took two parent chromosomes and mixed them together to get a single child by finding common features of both parents and passing them to the new chromosome.

The TSP has raised a lot of interest among researchers involved in the use of GAs as a tool for solving combinatorial optimization problems. Cheng and Gen (1997) listed a number of crossover techniques that fall within the canonical approach discussed by Holland et al. (1986) and that have been used in the solution of the TSP. They also reviewed a heuristic approach that is useful in many optimization problems. Within the canonical approach, Cheng and Gen (1997) mentioned various crossover methods such as partial-mapped crossover (PMX), a method originally proposed by Goldberg and Lingle (1987). Other methods mentioned are the order crossover (OX) method, proposed by Davis (1985), the position-based crossover method, proposed by Syswerda (Davis, 1985), and the cycle crossover (CX) method, proposed by Holland et al. (1986). Yamamura et al. (1992) developed a method specifically used in the TSP called subtour exchange crossover. The second approach, called the heuristic approach, is a very problem-dependant approach as different heuristics can be used depending on the combinatorial optimization problem being studied. Many authors, however, have used the heuristic crossover method presented by Grefenstette et al. (1987) for solving the TSP problem. This approach follows the nearest neighbor heuristic for finding the best path from one place to another. As mentioned before, crossover may be the most important feature in a GA. The performance of the algorithm when trying to solve a combinatorial optimization problem is closely related with the performance of the crossover operation. A very critical characteristic of the crossover operation is the crossover rate, which controls the number of chromosomes that undergo the crossover operation. This rate should be carefully defined so that an optimal solution can be found in a reasonable amount of time. A high crossover rate allows a more deep exploration of the solution space and reduces the chances of settling for a false optimum; but if the rate is too high, the result is wastage of computation time in exploring bad regions of the solution space. In the sample shown at the end of the section, the crossover rate used is 0.60, meaning that about 60% of the chromosomes experience crossover. Figure 11.2 summarizes the crossover methodologies.

Mutation is an operation that produces spontaneous random changes in some chromosomes. It serves one of two roles: replacing genes lost from the population during the selection process (so that they can be tried in a new context), or providing the genes that were not present in the initial population. There are two

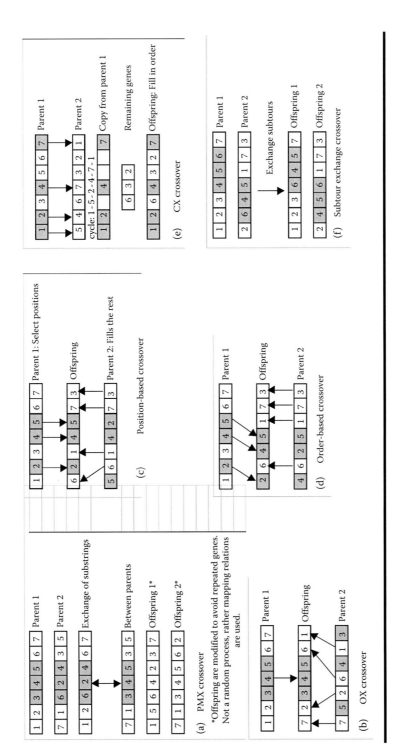

Figure 11.2   Crossover methodologies.

basic ways of doing mutation. They are bit-flipping and random assignment. In bit-flipping, bits in a chromosome are changed with a certain probability. Random assignment assigns zeros or ones at random within the chromosome disregarding the values that were already there. Falkenauer and Delchambre (1992) developed a mutation method that worked together with injection crossover for the solution of the knapsack problem. After crossover was performed, the mutation operation would delete a number of genes at random, and then it would append those same items in random order to arrive at new chromosomes. Tate and Smith (1995) developed a mutation method called inverse mutation, which worked together with the crossover operation used for the solution of the quadratic assignment problem. This method essentially took two genes at random and reversed the order of all sites within the subsequence bounded by the two selected genes. For the solution of the TSP, several mutation methods have been used and proposed during the last few years. Cheng and Gen (1997) summarized a number of mutation methods to solve this problem. These methods included the insertion mutation method, which selects a gene from a chromosome at random and inserts it in another place within the same chromosome; and the reciprocal exchange mutation method, which selects two positions within a chromosome at random and swaps their contents to produce new chromosomes. Cheng and Gen (1997) also mention displacement mutation, a method developed by Yamamura et al. (1992) that worked with their crossover operation in the solution of the TSP. Finally, Cheng and Gen (1997) introduced the heuristic mutation method, a technique developed by them in 1996. It uses the heuristic discussed in the previous section in order to produce improved offspring. It is important to mention again that these heuristics are very problem dependent, as each author develops his orher own heuristic for the combinatorial optimization problem they may be solving. Figure 11.3 summarizes the mutation methodologies.

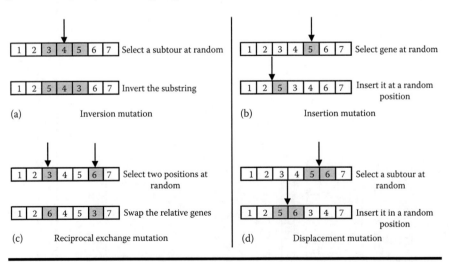

Figure 11.3   Mutation methodologies.

As for the crossover operation, a mutation rate has to be carefully defined for each problem. This mutation rate is defined as a percentage of the total number of genes in the population. This rate controls the pace at which new genes are introduced into the population for trial. If the rate is too low, many useful genes will never be tried out; but if the rate is too high, there will be much random perturbation and the offspring will start losing their resemblance to the parents. In the experiment performed at the end of the section, the mutation rate used is 0.15, meaning that about 15% of the total number of genes will experience mutation.

The principle behind GAs, as mentioned earlier, is Darwinian natural selection. Selection deals with the problem of selecting the "valuable" individuals or chromosomes that will survive and pass to the next generation. Selection is the driving force in any GA, and thus selection pressure is critical on the performance of a GA in the solution of a combinatorial optimization problem. Michalewicz (1992) stated that usually low selection pressure is indicated at the start of the GA search in favor of a wide exploration of the search space, while high selection pressure is recommended at the end in order to exploit the most promising regions in the search space. The selection procedure may create a new population for the next generation based on all parents and offspring or only on part of them. Selection may replace parent chromosomes by their offspring. Cheng and Gen (1997) explained that when all parents are replaced by their offspring, a generational replacement has taken place. This occurrence is not recommended and is highly undesirable, as GAs are blind in nature and so offspring chromosomes may be worst than their parents. This would lead the GA to some unwanted places when looking for an optimal solution. In the case of the TSP, a generational replacement would mean that all the plans and paths found by the algorithm would be thrown away and essentially, the algorithm would start looking for a solution from scratch. Of course, if one is lucky, and the new solutions are better, the algorithm will achieve its goal anyway and no effects will be perceived. However, the risk of a generational replacement should not be taken, especially if the combinatorial optimization problem being studied is rather large, as the costs associated with a generational replacement can be very high.

Over the past few years, some solutions have been suggested to overcome the problem of experiencing a generational replacement during the solution of a combinatorial optimization problem. Holland (1975) gave a first suggestion that involved a random selection of a parent whenever an offspring was born. This chosen parent was then replaced by the offspring that was just created. De Jong (1975) suggested another solution to the generational replacement problem. It is called the crowding strategy, which selects the parent to be replaced depending on its similarities with the just created offspring (the most similar parent dies and gives space for the new offspring). In the sample TSP, the sampling space is large and includes both parents and offspring. This means that both "old" and "new" solutions will fight for survival, and the idea of choosing a parent to be replaced by the offspring is not used. Instead, the fittest individuals will survive and will pass on to the next generation. The fitness of an individual or solution is calculated as the distance between the cities that produce the solution. The solutions that yield the shortest path or cost of travel between the cities will pass on to next generations, while the ones that yield a higher cost (and thus a longest path) will eventually die. After a few generations, the algorithm returns the chromosome with the lowest cost and distanced travel, which represents the best solution available to this combinatorial optimization problem. The selection of chromosomes is not completely determined by their fitness values. One advantage that GAs bring to the solution of various combinatorial optimization problems is the fact that they search the solution space more broadly, meaning that apparently "weak" solutions are not disregarded right away, as they may lead the algorithm to good final solutions. Holland (1975) proposed the most common method for the selection of chromosomes that pass from generation to generation. It is called the roulette wheel selection method, the most recognized method for selection among stochastic techniques.

The roulette wheel selection method determines the selection probability for each chromosome by proportionally assigning a portion of the roulette to a chromosome depending on its fitness value. After all probabilities have been found, roulette with the values is created. The selection process is then started by "spinning" the wheel as many times as required until a full population is selected. The probability that a chromosome will be chosen for survival is directly related to its fitness value, as chromosomes with higher fitness values occupy most of the roulette. In essence, chromosomes with high fitness values will be selected more times from this "biased" roulette.

The sample TSP uses this method for the selection of chromosomes. As mentioned earlier, the algorithm assigns fitness values to each solution by calculating the cost associated with the solution. This value essentially represents the distance traveled by the salesman if he or she were to take the path represented by the chromosome. Using these values, the algorithm assigns probabilities to each chromosome, and by spinning the roulette, the new generations are selected. The TSP will now be presented. As mentioned before, the GA used follows those techniques explained throughout this chapter.

11.4 Modular Systems Development

In general, modular systems can be developed by decomposing a system into its basic functional elements, mapping these elements into basic physical components and then integrating the basic components into a modular system capable of achieving the intended functions. This approach faces two important challenges (Pimmler and Eppinger, 1994): (1) Decomposition: Finding the most suitable set of subproblems may be difficult. (2) Integration: Combining the separate subsystems into an overall solution may also be difficult. To fully comprehend the underlying foundations of modular systems development, decomposition categories are further discussed.

11.4.1 Decomposition Categories

System decomposition is expected to result in two benefits (Pimmler and Eppinger, 1994): (1) Simplification: Decomposing large systems into smaller ones will lead to a reduction in the size of the problem that needs to be solved, which will make it easier to manage. (2) Speed: Solving smaller problems concurrently (parallel solutions) will reduce the time needed to solve the overall problem.

Decomposition methods can be categorized according to the area into which they are being applied: product decomposition, problem decomposition, and process decomposition (Kusiak and Larson, 1995).

11.4.1.1 Product Decomposition

Product decomposition can be performed at various stages of the design process and can be defined as the process of breaking the product down into physical elements from which a complete description of the product can be obtained. Two approaches are used in product decomposition: product modularity and structural decomposition.

11.4.1.1.1 Product Modularity

Product modularity is the identification of independent physical components that can be designed concurrently or replaced by predesigned components that have similar functional and physical characteristics. Product modularity relies on the lack of dependency between the physical components.

The computer industry provides an excellent example of modular products. The major components of a computer are shown in Figure 11.4. Major components are manufactured by different suppliers, which allow the manufacturers of microprocessors to choose from a wide library of products.

11.4.1.1.2 Structural Decomposition

The system is decomposed into subsystems, which are further decomposed into components leading to products, assemblies, subassemblies, and parts at the detailed design stage. The decomposition is represented in a hierarchy structure that captures the dependencies between subsystems. For example, the structural decomposition of a vehicle system is shown (Figure 11.5) and the carriage unit is further decomposed into components (Figure 11.6).

11.4.1.2 Problem Decomposition

For centuries, complex design problems were handled by breaking them into simpler, easy-to-handle subproblems. Problem decomposition should continue

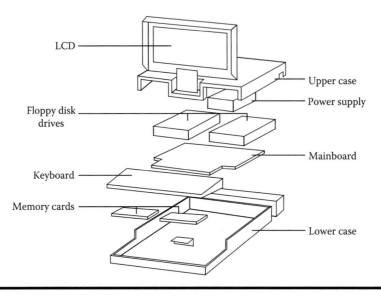

Figure 11.4 PC assembly diagram.

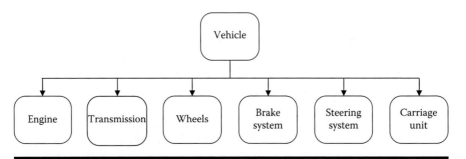

Figure 11.5 Structural decomposition of a vehicle system.

until basic independent products or units are reached. The interaction between the basic products should be identified and introduced as constraints imposed by higher subproblems. Problem decomposition is divided into requirements decomposition, constraint-parameter decomposition, and decomposition-based design optimization.

11.4.1.2.1 Requirements Decomposition

Requirements represent an abstraction of the design problem, starting with the overall requirement (general demand) and ending with the specific requirements (specific demands). The ability to meet a requirement is given by a design function. The requirements decompositions and their relationships to the corresponding

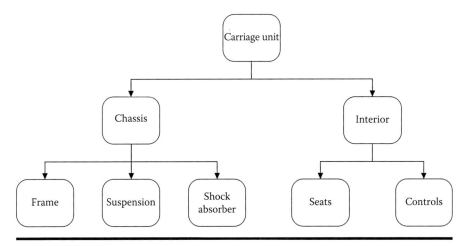

Figure 11.6 Structural decomposition of a carriage unit.

functions are represented in a tree diagram, where specific requirements are mapped into specific functions.

11.4.1.2.2 Constraint Parameter Decomposition

The parameters describe the features (quantitative or qualitative data) of the product, while the constraints define the ranges of values assigned to parameters that are defined by product requirements. The problem structure is represented in an incidence matrix (Kusiak and Larson, 1995). The incidence matrix is decomposed by grouping all nonempty elements in blocks at the diagonal. It is preferable that the blocks be mutually separable (independent). In some cases, overlapping between variables or constraints may occur. The design of a ball bearing is used to illustrate the decomposition (Kusiak and Larson, 1995). The parameters are listed in Table 11.1 and the constraints are shown in Table 11.2. The constraint-parameter incidence matrix is shown in Figure 11.7 and decomposed matrix is illustrated in Figure 11.8.

11.4.1.2.3 Decomposition-Based Design Optimization

The decomposition of a large complex design problem into smaller independent subproblems facilitates the use of mathematical programming techniques to solve and optimize the subproblems (Finger and Dixon, 1989a,b; Johnson and Benson, 1984). The solutions are integrated to provide an overall solution. The objective is to decompose a complex system into multilevel subsystems in a hierarchical form, in which a higher-level subsystem controls or coordinates the subsystems at the lower level. The subsystems are solved independently at the lower level. The objective at

Table 11.1   Ball Bearing Design Parameters

Parameter	Description
d_e	Pitch diameter
d_o	Outer-race diameter
d_i	Inner-race diameter
P_d	Diametral clearance
D	Rolling-element diameter
l	Race conformity ratio
R	Race curvature radius
B	Total conformity
l_o	Outer-race conformity
l_i	Inner-race conformity
D	Race curvature distance
β_l	Free contact angle
r_o	Outer-race curvature
r_i	Inner-race curvature
P_e	Free endplay
S	Shoulder height
θ	Shoulder angle height
R	Curvature sum
R_x	x direction effective radius
R_y	y direction effective radius
Γ	Curvature difference
β	Contact angle

Table 11.2   Ball Bearing Design Constraints

C_1	$d_e = \dfrac{1}{2}(d_o + d_i)$	C_7	$P_e = 2D \sin \beta_f$
C_2	$P_d = d_o - d_i - 2d$	C_8	$s = r(1 - \cos\theta)$
C_3	$f = r/d$	C_9	$\dfrac{1}{R} = \dfrac{1}{R_x} + \dfrac{1}{R_y}$
C_4	$B = f_o + f_i - 1$	C_{10}	$\Gamma = R\left(\dfrac{1}{R_x} - \dfrac{1}{R_y}\right)$
C_5	$D = Bd$	C_{11}	$R_x = d(d_e - d\cos\beta)/2d_e$
C_6	$\beta_f = \arccos \dfrac{r_o + r_i - \dfrac{1}{2}(d_o - d_i)}{r_o + r_i - d}$	C_{12}	$R_y = f_i d/(2f_i - 1)$

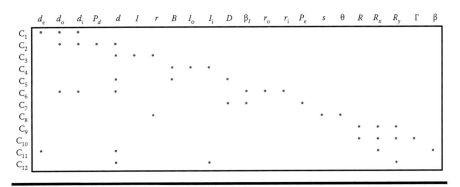

Figure 11.7   Ball bearing design constraint-parameter incidence matrix.

the higher level is to coordinate the action of the first level to ensure that the overall solution is obtained.

11.4.1.3 Process Decomposition

Process decomposition is the decomposition of the entire design process, starting with the need recognition and ending with the detail design. The activities in the design process are modeled in a generic manner independent of the specific product being designed. Three perspectives of process decomposition were recognized.

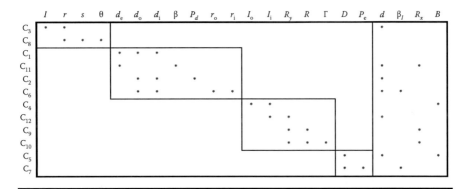

Figure 11.8   Decomposed constraint-parameter incidence matrix.

These are product flow perspective, information flow perspective, and resource perspective.

11.4.1.3.1 Product Flow Perspective

Design activities required to translate customer requirements into a detailed design of products are the focus of this perspective. The design activities are modeled as blocks with identified inputs and outputs (the output of one activity becomes the input of another activity). The decomposition tries to eliminate redundant activities and reorganize other activities to be performed concurrently, which will eventually reduce the product development time.

11.4.1.3.2 Information Flow Perspective

Analysis of the precedence constraints between the design activities is the main concern of this perspective. Precedence constraints are utilized to generate the required information needed to build supporting databases and communication networks and to schedule design activities, all concurrently.

11.4.1.3.3 Resource Perspective

The resources provide activities with a mechanism for transforming inputs to outputs. In this perspective, two types of constraints are considered:

1. External resource constraints, in which the resource used by the activity is generated by an activity or resource that is external to the design process.
2. Internal resource precedence constraints, in which the resource is developed in the design process and used by other activities.

11.4.1.3.4 Genetic Algorithms and Modular Design

Throughout the previous section, the need of developing an efficient algorithm that will produce acceptable answers for a wide variety of combinatorial optimization problems has been stressed. The algorithm developed can be used to solve a vast variety of combinatorial optimization problems, as the differences among these problems lie only on the chromosome representation of the answers. The proposed algorithm consists of the following steps (Kamrani and Salhieh, 2000; Gonzalez, 2001):

1. Initialization:
   a. Define the chromosomes and their representation.
   b. Determine the population size, crossover probability, and mutation probability.
   c. Generate initial population of solutions that satisfy the problem constraints.
2. Reproduction:
   a. Compute the fitness value of each chromosome.
   b. Calculate the total fitness of the population.
   c. Find the reproduction probability for each chromosome.
   d. Calculate the cumulative reproduction probability for each chromosome.
   e. Choose the best chromosomes for the next generation.
3. Crossover:
   a. Do crossover on the selected chromosomes based on a crossover probability defined in the initialization procedure.
4. Mutation:
   a. Perform mutation on the chromosomes based on a mutation probability defined in the initialization procedure.

The first step in the implementation of the algorithm involves the representation of the problem to be solved with a finite-length string called chromosome. Each element of the chromosome (each gene) represents a decision variable, feature, or parameter of the problem. The chromosome representation involves two related decisions. The first decision has to do with the problem of finding an effective way of encoding the problem in terms of a string chromosome. This decision can lead to the determination of the size of each chromosome, which is completely dependent on the nature of the problem. The second decision involves the selection of a string format for each gene in the chromosome. The value of a gene can be either a binary number or an integer. The binary format has been widely used. The integer format is domain specific and has also been widely used in the areas such as facility layout, TSPs, scheduling, among many other problems. Each gene will have an integer value, which will represent the classification of subsystem/components into groups depending on their similarities. Encoding for this problem is done as follows: the chromosome strings consist of n integer genes, each of which represent a component in the family of components $(1, \ldots, n)$. Each value in a chromosome stands for the module number to which a subsystem/component is assigned. For example, if there are 10 parts to be divided into 3 modules, then the chromosome can be represented as $(1, 1, 3, 2, 1, 2, 2, 3, 3, 3)$. Here, this chromosome indicates that module #1 has parts 1, 2, and 5; module #2 has parts 4, 6, and 7; and module #3 has parts 3, 8, 9, and 10. The size of the chromosome depends only on the size of the matrix being studied, so if a problem has 20 parts/components, the size of the chromosome will be 20 (20 genes). Also, since the similarities between components can and will be different, the integer value of each gene is allowed to be any number between 1 and the number of components being grouped (n), which will basically mean that any group from 1 to n could be selected in the process.

Some parameters, including population size, maximum number of generations, and the probability of crossover and mutation, should be decided before the algorithm starts to find a solution. These parameters are very sensitive to the computational performance of the algorithm. The population size directly affects the computation time of the algorithm, as having a large population size will mean that the algorithm will have to perform more calculations. It is important to note that a higher population size will give the algorithm a higher chance of success, as the solution space searched will be larger. However, if good solutions are required in a reasonable amount of time, the population size should be smaller. This will allow the algorithm to quickly find a good solution to the problem being studied. If the crossover probability is set to be .85, this basically means that about 85% of the chromosomes in each generation will experience crossover. If the crossover probability is set to 100%, then all offspring will be created by crossover. If it is 0%, the whole new generation will be made from exact copies of chromosomes from old populations. It is acceptable to allow some chromosomes to survive without change in the next generation, an occurrence called elitism. If the mutation probability is set to be .15, this means that about 15% of the genes in a population will experience mutation. If the mutation probability is set to a 100%, all of the chromosome will be changed whereas if it is set to 0%, nothing will change. Mutation occurs to prevent falling into local optima, but it should not occur very often, as the GA will become just a random search method.

The number of generations is directly affected by the performance of the algorithm. A maximum number of generations can be set, but the algorithm will usually find a solution to the problem before this number is ever reached. This means that many extra computations will be performed after the algorithm has already found a solution. The algorithm developed examines the fitness of each population, and stops whenever the global fitness cannot be greater than what it already is. This means that the algorithm has found an optimal solution, at which point, the solution is given to the user while the algorithm starts another run to investigate the grouping of components with a different number of groups. After the problem presentation, a set of initial solutions (chromosomes) called population has to be determined. This initial population will represent the first generation, which will then evolve until a good solution to the problem is determined. In order to obtain feasible solutions, each chromosome has to satisfy the problem constraints. The replacement policy is a way of making sure that the chromosomes created for the first generation are valid. Validity can be asserted in many ways, but one easy method checks the problem constraints given by the problem with each created chromosome. Two different methods have been widely used. The first method is called the variable restriction method, which only chooses the chromosomes that meet the feasible region of constraints. The second method is called the penalty function method, which allows chromosomes to violate the constraints by giving them a penalty, which will then in turn lower their probability of survival to zero.

In the proposed method, a modification of the variable restriction is utilized. A random number generator creates the initial population by randomly generating numbers between 1 and n (where n is the size of the chromosome). After each chromosome is created, the algorithm checks which groups were created, and with this information, the algorithm checks the validity of the chromosome by comparing it with the specified problem constraints. If the chromosome is valid, no changes are made, but if it violates one of the constraints specified in the problem, the necessary changes are made to ensure the validity of the chromosome. Due to the chromosomal representation chosen for the problem, the only problem that can occur is the assignment of subsystems/components to modules that have not been created. The problem is easily resolved by creating the modules that are represented within the chromosome. For example, if the modules selected are 2, 5, and 8, and the chromosome is (5, **2**, 8, 8, **2**, 5, 5, **8**, 2, 2), one has to make sure that position 2 in the chromosome has a 2, position 5 in the chromosome has a 5, and position 8 in the chromosome has an 8. In this case, the chromosome would not be valid, and the algorithm would change the fifth position value from a 2 to a 5 to make the chromosome a valid one. This modification of the variable restriction method is used after each population is created, so that only valid chromosomes are manipulated.

A fitness function is required for evaluating and selecting good generations of chromosomes. The fitness function should give domain-specific information about the value of each chromosome. For this reason, it is usually a good idea to define it in the form of a mathematical formulation, either a maximization or a minimization of some parameters and constraints. In the case of this model, the objective function of the optimization model is selected as the fitness function. Each chromosome, created by either the initialization of the GA, or by the creation of a new generation, will have a fitness value. Since the objective of this model is to maximize the sum of the similarities, the fitness values are continuously increasing, until an optimal solution is found. Once the fitness function is defined and used for the first time (after the first generation of chromosomes are created), the algorithm starts the selection and reproduction process. Using the fitness values of each chromosome, the roulette wheel selection process begins by finding the total fitness of the population. Using the roulette principles, a reproduction probability is assigned to each chromosome (based on their fitness values), and the roulette is filled using the respective cumulative probabilities of each chromosome. Since the space on the wheel is totally dependent on the fitness value of each chromosome, fitter chromosomes will have a larger space in this bias roulette, increasing their chances of survival. The algorithm then randomly generates a number between 0 and 1, and a chromosome is chosen as a parent based on that random number and the cumulative reproduction probability. At the end of the reproduction cycle, the new group of chromosomes is ready for crossover and mutation. After the better-fitted chromosomes are selected, each pair of chromosomes is selected sequentially to exchange information according to the crossover probability previously determined. A random number is generated, and if the number lies below the crossover probability, crossover is performed. A mode of single-point crossover is used in this method, a mode for which the cut point for doing crossover is randomly selected at each instance. All but one chromosome are subject to crossover. After reproduction gives a new generation of chromosomes, the best chromosome is assured to be a part of the next generation unchanged, as it is always a good idea to keep the best solution within a population. Elitism ensures that the best chromosome will not suffer crossover and/or mutation. The algorithm will also benefit from elitism.

After all chromosomes have been exposed to crossover, the new generation is subjected to mutation. A crossover cycle is over whenever each pair of sequential chromosomes have either experienced crossover (because their probability of crossover fell within the P_{cr}%), or have been passed without change because their probabilities fell outside the crossover probability. If a pair of chromosomes experience crossover, the next chromosomes to be sequentially selected will not include any of these two chromosomes, as it is not desired to double-crossover a single chromosome. Mutation in each chromosome occurs according to previously set mutation probability. The mutation operator occasionally alters genes within a chromosome by changing the value of a single gene within a chromosome. The GA in this case alters the genes based on the mutation probability as well as on the groups or modules that have already been selected for each chromosome. This ensures that new chromosomes will be explored, and thus, the solution space will be broader. Once all chromosomes are subjected to mutation (except the best one), the new generation of chromosomes is set. The new run will not require an initialization procedure, as the generation that just experienced reproduction, crossover, and mutation, acts as the new generation. However, another mode of mutation is introduced at this point, as the chromosomes of this new generation are subjected to a validity check. An illegal chromosome should never be passed to the next generation (replacement policy), and so, the required changes to each chromosome are made in order to ensure their validity.

For example, similarity matrix, which is used to measure the degree of association between the subsystem/components of a proposed system, is illustrated in Figure 11.9.

The GA begins by randomly generating the initial population of solutions for the similarity matrix, since the first generation of solutions group all the elements into

	Gear 1	Gear 2	Gear 3	Gear 4	Shaft 1	Shaft 2	Shaft 3	Bearing 1	Bearing 2	Bearing 3	Bearing 4	Bearing 5	Bearing 6	Key 1	Key 2	Key 3	Key 4
Gear 1		5	1	1	6	1	1	2	2	1	1	1	1	6	1	1	1
Gear 2	5		2	1	1	6	1	1	1	2	2	1	1	1	6	3	2
Gear 3	1	2		2	1	6	1	1	1	2	2	1	1	1	1	6	1
Gear 4	1	1	2		1	1	6	1	1	1	1	5	5	1	1	1	6
Shaft 1	6	1	1	1		1	1	5	5	1	1	1	1	6	1	1	1
Shaft 2	1	6	6	1	1		1	1	1	5	5	1	1	1	6	6	1
Shaft 3	1	1	1	6	1	1		1	1	1	1	5	5	1	1	1	6
Bearing 1	2	1	1	1	5	1	1		2	1	1	1	1	2	1	1	1
Bearing 2	2	1	1	1	5	1	1	2		1	1	1	1	2	1	1	1
Bearing 3	1	2	2	1	1	5	1	1	1		2	1	1	1	2	2	1
Bearing 4	1	2	2	1	1	5	1	1	1	2		1	1	1	2	2	1
Bearing 5	1	1	1	5	1	1	5	1	1	1	1		2	1	1	1	2
Bearing 6	1	1	1	5	1	1	5	1	1	1	1	2		1	1	1	2
Key 1	6	1	1	1	6	1	1	2	2	1	1	1	1		1	1	1
Key 2	1	6	1	1	1	6	1	1	1	2	2	1	1	1		2	1
Key 3	1	3	6	1	1	6	1	1	1	2	2	1	1	1	2		1
Key 4	1	2	1	6	1	1	6	1	1	1	1	2	2	1	1	1	

Figure 11.9 Similarity matrix.

one group. The fitness value for the best chromosome is 48. The grouping for this case does not give the user any valuable information as all the subsystem/components are grouped into one family. After 8 generations, the algorithm reaches the best solution for this case, a solution with a total fitness (for the generation) of 960, which is the sum of all the fitness values of each chromosome in a generation composed of 20 individuals. A maximum of 80 generations is allowed, but in this case, the algorithm only needs 8 generations mainly because of the simplicity of the problem being studied (classification into one module). The second run groups the subsystem/components into two separate modules. At the end of this run, the fitness values of each chromosome should be greater than the ones given in the previous run. However, if the groupings do not yield a better fitness value, this will mean that grouping the subsystem/components into two groups does not give the best overall optimal solution. It is important to note that some groupings, even if they are not the optimal ones, will give companies what they are looking for, as they may be satisfied with good solutions that will improve their operations. The fitness function value is 56, and the best grouping is (4, 6, 6, 4, 4, 6, 4, 6, 6, 6, 6, 4, 4, 6, 6, 6, 4). This basically means that, for example, subsystem/components 1 and 4 are together in one module (positions one and four are both fours), and subsystem/components 2 and 3 are together in another module. A total of 75 generations were needed. The GA solution for two modules is a good solution, but it is not the optimal one, as there are several high similarities left out of the two groups (an occurrence that may imply that the solution is not optimal). It is apparent from the grouping that the addition of at least another group would improve the arrangement of subsystem/components, as the assignment of the subsystem/components into two groups can be very restrictive. Figure 11.10 shows the final groupings.

The third run of the GA groups the subsystem/components into a total of three modules. This new run of the GA brings a new improvement with respect to the previous runs. In this case, the GA yields a fitness value of 64, the highest one so

	SS1C1	SS1C4	SS2C1	SS2C3	SS3C5	SS3C6	SS4C4	SS1C2	SS1C3	SS2C2	SS3C1	SS3C2	SS3C3	SS3C4	SS4C1	SS4C2	SS4C3
SS1C1		2	5	2	2	2	2	6	2	2	1	1	2	2	5	2	2
SS1C4	2		2	5	4	4	5	2	1	2	2	2	2	2	2	2	2
SS2C1	5	2		2	2	2	2	2	2	2	4	4	2	2	5	2	2
SS2C3	2	5	2		4	4	5	2	2	2	2	2	2	2	2	2	2
SS3C5	2	4	2	4		1	1	2	2	2	2	2	2	2	2	2	2
SS3C6	2	4	2	4	1		1	2	2	2	2	2	2	2	2	2	2
SS4C4	2	5	2	5	1	1		3	2	2	2	2	2	2	2	2	2
SS1C2	6	2	2	2	2	2	3		1	5	2	2	1	1	2	5	2
SS1C3	2	1	2	2	2	2	2	1		5	2	2	1	1	2	2	5
SS2C2	2	2	2	2	2	2	2	5	5		2	2	4	4	5	5	5
SS3C1	1	2	4	2	2	2	2	2	2	2		1	2	2	1	2	2
SS3C2	1	2	4	2	2	2	2	2	2	2	1		2	2	1	2	2
SS3C3	2	2	2	2	2	2	2	1	1	4	2	2		1	2	1	1
SS3C4	2	2	2	2	2	2	2	1	1	4	2	2	1		2	1	1
SS4C1	5	2	5	2	2	2	2	2	2	2	1	1	2	2		2	2
SS4C2	2	2	2	2	2	2	2	5	2	5	2	2	1	1	2		1
SS4C3	2	2	2	2	2	2	2	2	5	5	2	2	1	1	2	1	

Figure 11.10 Grouping for a two modules.

	SS1C1	SS2C1	SS3C1	SS3C2	SS4C1	SS1C2	SS1C3	SS2C2	SS3C3	SS3C4	SS4C2	SS4C3	SS1C4	SS2C3	SS3C5	SS3C6	SS4C4
SS1C1		5	1	1	5	6	2	2	2	2	2	2	2	2	2	2	2
SS2C1	5		4	4	5	2	2	2	2	2	2	2	2	2	2	2	2
SS3C1	1	4		1	1	2	2	2	2	2	2	2	2	2	2	2	2
SS3C2	1	4	1		1	2	2	2	2	2	2	2	2	2	2	2	2
SS4C1	5	5	1	1		2	2	2	2	2	2	2	2	2	2	2	2
SS1C2	6	2	2	2	2		1	5	1	1	5	2	2	2	2	2	3
SS1C3	2	2	2	2	2	1		5	1	1	2	5	1	2	2	2	2
SS2C2	2	2	2	2	2	5	5		4	4	5	5	2	2	2	2	2
SS3C3	2	2	2	2	2	1	1	4		1	1	1	2	2	2	2	2
SS3C4	2	2	2	2	2	1	1	4	1		1	1	2	2	2	2	2
SS4C2	2	2	2	2	2	5	2	5	1	1		1	2	2	2	2	2
SS4C3	2	2	2	2	2	2	5	5	1	1	1		2	2	2	2	2
SS1C4	2	2	2	2	2	2	1	2	2	2	2	2		5	4	4	5
SS2C3	2	2	2	2	2	2	2	2	2	2	2	2	5		4	4	5
SS3C5	2	2	2	2	2	2	2	2	2	2	2	2	4	4		1	1
SS3C6	2	2	2	2	2	2	2	2	2	2	2	2	4	4	1		1
SS4C4	2	2	2	2	2	3	2	2	2	2	2	2	5	5	1	1	

Figure 11.11   Grouping for a three-group solution.

far, and a total fitness of 1263. This run uses 80 generations to get the optimal solution for the problem, a solution that will end up being the overall optimal solution for the speed reducer. The result of this run is shown in Figure 11.11.

The grouping given by the GA is (5, 6, 6, 7, 5, 6, 7, 5, 5, 6, 6, 7, 7, 5, 6, 6, 7). The solution for the three-module classification is, in fact, the optimal solution. The total fitness (as well as the fitness of the chromosomes) of the new run for four-module grouping goes down to almost 900, while for the previous run it was over 1200. Moreover, the fitness value of the best chromosome goes down from 64 to just 47, which is actually lower than the fitness value of the first run, in which the algorithm grouped all elements into one group. This sudden drop in fitness, along with the poor grouping of the subsystem/components (1, 1, 6, 1, 5, 6, 5, 5, 5, 6, 6, 15, 5, 5, 15, 6, 15), gives the user a "preview" of what is about to happen. Figure 11.12 lists the solution for the four-module classification.

	SS1C1	SS1C2	SS1C4	SS1C3	SS2C2	SS3C3	SS3C4	SS4C3	SS2C1	SS2C3	SS3C1	SS3C2	SS3C6	SS4C1	SS3C5	SS4C2	SS4C4
SS1C1		6	2	2	2	2	2	2	5	2	1	1	2	5	2	2	2
SS1C2	6		2	1	5	1	1	2	2	2	2	2	2	2	2	5	3
SS1C4	2	2		1	2	2	2	2	1	5	2	2	4	2	4	2	5
SS1C3	2	1	1		5	1	1	5	2	2	2	2	2	2	2	2	2
SS2C2	2	5	2	5		4	4	5	2	2	2	2	2	2	2	5	2
SS3C3	2	1	2	1	4		1	1	2	2	2	2	2	2	2	1	2
SS3C4	2	1	2	1	4	1		1	2	2	2	2	2	2	2	1	2
SS4C3	2	2	2	5	5	1	1		2	2	2	2	2	2	2	1	2
SS2C1	5	2	1	2	2	2	2	2		2	4	4	2	5	2	2	2
SS2C3	2	2	5	2	2	2	2	2	2		2	2	4	2	4	2	5
SS3C1	1	2	2	2	2	2	2	2	4	2		1	2	1	2	2	2
SS3C2	1	2	2	2	2	2	2	2	4	2	1		2	1	2	2	2
SS3C6	2	2	4	2	2	2	2	2	2	4	2	2		2	1	2	1
SS4C1	5	2	2	2	2	2	2	2	5	2	1	1	2		2	2	2
SS3C5	2	2	4	2	2	2	2	2	2	4	2	2	1	2		2	1
SS4C2	2	5	2	2	5	1	1	1	2	2	2	2	2	2	2		2
SS4C4	2	3	5	2	2	2	2	2	2	5	2	2	1	2	1	2	

Figure 11.12   Grouping for a four-module solution.

The grouping shown is a more specific grouping of the subsystem/components. However, as it can be seen on the matrix, there are high similarity values outside the groupings. This implies that the solution is not optimal, although it may give the user a much more specific arrangement of subsystem/components.

Author

Ali K. Kamrani is an associate professor of industrial engineering, and director of the industrial engineering graduate program studies and accelerated BS to MS programs. He is also the founding director of the Design and Free Form Fabrication Laboratory at the University of Houston, United States. He has been a visiting professor at the Princess Fatimah Alnijris's Research Chair for AMT, Industrial Engineering Department at King Saud University, Riyadh. Saudi Arabia. He received his BS in electrical engineering in 1984, his MEng in electrical engineering in 1985, his MEng in computer science and engineering mathematics in 1987, and his PhD in industrial engineering in 1991, all from the University of Louisville, Louisville, Kentucky. His research interests include the fundamental application of systems engineering and its application in the design and development of complex systems. He is the editor in chief for the *International Journal of Collaborative Enterprise* and the *International Journal of Rapid Manufacturing*. He is a professional engineer at the State of Texas.

References

Ahuja, R.K., Orlin, J.B. and A.Tiwari. 1999. A greedy genetic algorithm for quadratic assignment problem. *Computers & Operations Research* 27: 917–934.

Angel, E. and V. Zissimopoulos. 2000. On the classification of NP-complete problems in terms of their correlation coefficient. *Discrete Applied Mathematics* 99: 261–277.

Breedam, A.V. 1999. Comparing descent heuristics and metaheuristics for the vehicle routing problem. *Computers & Operations Research* 28: 289–315.

Cheng, R. and M. Gen. 1997. *Genetic Algorithms & Engineering Design*. New York: Wiley-Interscience Publication.

Chu, C.H., Premkumar, G., and H. Chou. 2000. Digital data networks using genetic algorithms. *European Journal of Operational Research* 127: 140–158.

Davis, L. 1985. Applying adaptive algorithms to domains. In: *Proceedings of the International Joint Conference on Artificial Intelligence*, Los Angeles, CA, pp. 162–164.

De Jong, K. 1975. An analysis of the behavior of a class of genetic adaptive systems. PhD dissertation. University of Michigan, Ann Arbor, MI.

Dellaert, N., Jeunet, J., and N. Jonard. 2000. A genetic algorithm to solve the general multi-level lot-sizing problem with time-varying costs. *International Journal of Production Economics* 68: 241–257.

Falkenauer, E. and A. Delchambre. 1992. A genetic algorithm for bin packing and line balancing. In: *Proceedings of the IEEE International Conference on Robotics and Automation*, Nice, France, pp. 1186–1193.

Finger, S. and J.R. Dixon. 1989a. A review of research in mechanical engineering design, Part I: Descriptive, prescriptive, and computer-based models of design processes. *Research in Engineering Design* 1: 51–67.

Finger, S. and J.R. Dixon. 1989b. A review of research in mechanical engineering design, Part II: Representation, analysis, and design for the life cycle. *Research in Engineering Design* 1: 121–137.

Goldberg, D. 1989. *Genetic Algorithms in Search, Optimization and Machine Learning.* Boston, MA: Addison-Wesley.

Goldberg, D. and R. Lingle. 1987. Alleles, loci and the traveling salesman problem. In: *Proceedings of the First International Conference on Genetic Algorithms*. Hillsdale, NJ: Lawrence Erlbaum Associates.

Gonzalez, R. 2001. A genetic algorithm based methodology for solving combinatorial optimization problems. Master thesis, University of Michigan, Dearborn, MI.

Gordon, V. and D. Whitley. 1992. Serial and parallel genetic algorithms as functions optimizers. In: *Proceedings of the Fifth International Conference on Genetic Algorithms*, San Francisco, CA, pp. 177–183.

Grefenstette, J., Gopal, R., Rosmaita, B.J., and D. Van Gucht. 1987. Genetic algorithms for the traveling salesman problem. In: *Proceedings of the First International Conference on Genetic Algorithms*. Hillsdale, NJ: Lawrence Erlbaum Associates, pp. 160–168.

Hoffman, K.L. 2000. Combinatorial optimization: Current successes and directions for the future. *Journal of Computational and Applied Mathematics* 124: 341–360.

Holland, J.H. 1975. *Adaptation in Natural and Artificial Systems.* Boston, MA: MIT press.

Ishibuchi, H., Nozaki, K., Yamamoto, N., and H. Tanaka. 1995. Selecting fuzzy if-then rules for classification problems using genetic algorithms. *IEEE Transactions on Fuzzy Systems* 3: 260–270.

Johnson, R.C. and R.C. Benson. 1984. A basic two-stage decomposition strategy for design optimization. *Transactions of the ASME* 106: 380–386.

Kamrani A. and S. Salhieh. 2000. *Product Design for Modularity.* New York: Kluwer Academic Publishers.

Kusiak, A. and N. Larson. 1995. Decomposition and representation methods in mechanical design. *Transactions of the ASME* 117B: 17–24.

Law, A.G. 1976. Theory of approximation, with applications. *Proceedings of a Conference Conducted by the University of Calgary and the University of Regina*, Alberta, Canada.

Mans, B., Mautor, T., and C. Roucairol. 1995. A parallel depth first search branch and bound algorithm for the quadratic assignment problem. *European Journal of Operational Research* 81: 617–628.

Michalewicz, Z. 1992. *Genetic Algorithms + Data Structures = Evolution Programs.* Berlin Heidelberg: Springer-Verlag.

Min, L. and W. Cheng. 1999. A genetic algorithm for minimizing the makespan in the case of scheduling identical parallel machines. *Artificial Intelligence in Engineering* 13: 399–403.

Obitko, M. 1998. *Introduction to Genetic Algorithms.* Prague, Czech Republic: Czech Technical University. http://cs.felk.cvut.cz/~xobitko/ga

Oliver, I.M., Smith, D.J., and J.R.C. Holland. 1987. A study of permutation crossover operators on the travelling salesman problem. *Proceedings of the Second International Conference on Genetic Algorithms*, pp. 224–230.

Olsen, A. 1993. Penalty functions and the knapsack problem. In: *Proceedings of the First IEEE Conference on Evolutionary Computation*. Orlando, FL: IEEE Press, pp. 554–558.

Pemberton, J.C. and Z. Weixiong. 1996. Epsilon-transformation: Exploiting phase transitions to solve combinatorial optimization problems. *Artificial Intelligence* 81: 297–325.

Pimmler, T.U. and S.D. Eppinger. 1994. Integration analysis of product decompositions. In: *ASME Design Theory and Methodology Conference*, Minneapolis, MN, p. 68.

Rios, R.Z. and J.F. Bard. 1998. Heuristics for the flow line problem with setup costs. *European Journal of Operational Research* 110: 76–98.

Sait, S. and H. Youssef. 1999. *Iterative Computer Algorithms with Applications in Engineering: Solving Combinatorial Optimization Problems*. Washington, DC: IEEE Computer Society.

Sridhar, R. and N. Chandrasekharan. 1995. Highly parallelizable problems on sorted intervals. *Parallel Computing* 21: 433–446.

Tate, D. and A. Smith. 1995. A genetic approach to the quadratic assignment problem. *Computers and Operations Research* 22: 73–83.

Tian, P., Ma, J., and D.M. Zhang. 1999. Application of the simulated annealing algorithm to the combinatorial optimization problem with permutation property: An investigation of generation mechanism. *European Journal of Operational Research* 118: 81–94.

Voudouris, C. and E. Tsang. 1999. Guided local search and its application to the traveling salesman problem. *European Journal of Operational Research* 113: 469–499.

Wall, M. 1995. *Introduction to Genetic Algorithms*. Boston, MA: Massachusetts Institute of Technology. http://lancet.mit.edu/~mbwall/presentations/IntroToGAs

Wang, R. 1998. *A Genetic Algorithm Based Knowledge Acquisition System for Equipment Diagnosis*. Dearborn, MI: University of Michigan-Dearborn.

Yamamura, M., Ono, T., and S. Kobayashi. 1992. Character-preserving genetic algorithms for the traveling salesman problem. *Journal of Japan Society for Artificial Intelligence* 6: 1049–1059.

Zhang, W. 1999. *State Space Search: Algorithms, Complexity, Extensions, and Applications*. Springer-Verlag, New York.

Index

T - #0011 - 160425 - C0 - 234/156/20 [22] - CB - 9781439809266 - Gloss Lamination